Patanjali's Yoga Philosophy

This text on Yoga Philosophy of Patanjali is based on his *Yoga Sutras*. In the *Yoga Sutras* Patanjali brought together various principles and practices of Yoga prevalent at his time into a coherent system. The *sutras* are short forms with a few essential words each. The *sutras*, therefore, need extensive explanations in order to understand their implications.

This text presents the teachings of Patanjali in some logical order in three Sections. In the first section, a brief Introduction to Yoga and its kinds, and the Yoga Tradition in India are presented. The Yoga Tradition covers the origin of *Yoga*, Yoga in the *Vedas,* Yoga in the *Upanishads*, Yoga in Epics, Yoga in the *Bhagavad Gita*, Patanjali's contribution to Yoga, Yoga in the *Srimad Bhagavatam, Puranas, Dharma Sastras, Agamas, Tirumantiram* and *Samhitas*.

Next, an Overview of Indian Philosophy, and the Samkhya Philosophy and Patanjali's Yoga Philosophy are briefly presented.

The second section is devoted for the discussion of Patanjali's Yoga Psychology. It comprises an analysis of the mind, mental modifications of the mind and their control, three *Gunas, Prakrti* and *Purusha*, God, afflictions: Causes and Remedy, and the Doctrine of Karma.

In the third section, Patanjali's Yoga system is explained. First, the preparatory practice of Kriya Yoga is outlined. Then an overview of the Raja Yoga is presented. After discussing the obstacles to Yoga, each of the eight limbs of Yoga is explained. Finally the Doctrine of Liberation is presented.

Patanjali's Yoga Philosophy

Based on the Teachings
of
Sri Swami Satchidananda

O.R. Krishnaswami

DEV PUBLISHERS & DISTRIBUTORS
New Delhi

Published by:
DEV PUBLISHERS & DISTRIBUTORS
2nd Floor, Prakash Deep,
4735/22, Ansari Road,
Darya Ganj,
New Delhi-110002
Phone : 011-43572647, 9810236140
e-mail: devbooks@hotmail.com
website: www.devbooks.co.in

NOTE

In this Text the words 'man,' 'he,' and 'him' are used in a generic sense and refer to both men and women.

ISBN 978–81–920752-3-5
First published 2011

Printed in India

Contents

Preface

This text on Yoga Philosophy of Patanjali is based on his *Yoga Sutras*. In the *Yoga Sutras* Patanjali brought together various principles and practices of Yoga prevalent at his time into a coherent system. The *sutras* are short forms with a few essential words each. The *sutras*, therefore, need extensive explanations in order to understand their implications.

Some commentaries listed under references were consulted for writing this work. I acknowledge my grateful indebtedness to their authors.

This text presents the teachings of Patanjali in some logical order in three Sections. In the first section, a brief Introduction to Yoga and its kinds, and the Yoga Tradition in India are presented. The Yoga Tradition covers the origin of *Yoga*, Yoga in the *Vedas,* Yoga in the *Upanishads*, Yoga in Epics, Yoga in the *Bhagavad Gita*, Patanjali's contribution to Yoga, Yoga in the *Srimad Bhagavatam, Puranas, Dharma Sastras, Agamas, Tirumantiram* and *Samhitas*.

Next, an Overview of Indian Philosophy, and the Samkhya Philosophy and Patanjali's Yoga Philosophy are briefly presented.

The second section is devoted for the discussion of Patanjali's Yoga Psychology. It comprises an analysis of the mind, mental modifications of the mind and their control, three *Gunas, Prakrti* and *Purusha*, God, afflictions: Causes and Remedy, and the Doctrine of Karma.

In the third section, Patanjali's Yoga system is explained. First, the preparatory practice of Kriya Yoga is outlined. Then an overview of the Raja Yoga is presented. After discussing the obstacles to

Yoga, each of the eight limbs of Yoga is explained. Finally the Doctrine of Liberation is presented.

The writing of this text was entirely due to the Grace of the Supreme Lord, and the Grace of my Gurudev Sri Swami Satchidananda, the founder of Satchidananda Ashram and the Integral Yoga Institute in Virginia with centers in several countries. This humble soul was a mere instrument in their hands. With deep devotion this text is dedicated to the Lotus Feet of the Lord and Sri Gurudev.

May this work inspire and guide the students of Yoga in their study and practice for achieving the ultimate purpose of life, God-realization.

O. R. KRISHNASWAMI

Acknowledgement

I am deeply grateful to Professor Kuldip C. Gupta, Ph.D., the President of the Hindu University of America, Orlando, Florida for having encouraged me to write this text for the benefit of his students.

I am indebted to the authors of the books listed under references and cited in the footnotes, from whose thoughts, I derived enlightenment.

I am thankful to my sons and daughter who presented me with a laptop computer and also procured several books for my study and reference.

I am also thankful to my third son's family and my family, which gave love, encouragement and support for doing this work.

O. R. KRISHNASWAMI

May 2, 2010
New Delhi (India)

SECTION I

Introduction

1.1: What is Yoga?

Introduction

Yoga is a spiritual science. It is concerned with man's quest for realizing his true nature. He is essentially divine, perfect and infinite. He is an embodied soul or Self or Atman. The Atman is eternal, pure and perfect. However, as it is embodied in a human form with a body, mind and senses, the apparent person (or the lower self) is unaware of his real nature. He falsely identifies himself with his body-mind system, which is a part of *Prakrti* (Nature) composed of matter and therefore, it is subject to decay and death. Driven by the forces of Nature, called three *Gunas*, he seeks happiness from worldly objects and sense enjoyments; and falls a prey to endless desires. He binds himself by the effects of his selfish actions, for every action there is an equal reaction. Yoga helps the apparent person to cast off his ignorance and enable him to realize his true nature, viz., the eternal Self or Spirit. With such realization Liberation is attained from sorrows and sufferings and from the bondage of the effects of actions (Karma). **Yoga** is, thus, a **means for Self-realization and Liberation.** Through the constant practice of Yoga in a spirit of dispassion, the thought-waves of the mind are stopped, the mind becomes pure and focused on the higher self. Then the Yogi is established in his own nature, i.e., he attains Self-realization.

Definition of Yoga

The Sanskrit word "*Yoga*" is derived from its root "*yuj*" which means "to join" or "to unite." What are the two things that join together? *Ajivatman* (embodied soul) unites with the *Paramatman*

(Supreme Self or Spirit). Although the *jivatman* is a facet of the *Paramatman*, and in essence both are the same, the *jivatman* has become subjectively separated from *Paramatman* or God. However, after going through an evolutionary cycle in the manifested universe, it becomes united with *Paramatman*. This **union** of *jivatman* with *Paramatman*, and the **methods** by which the union is attained, both are called Yoga. Yet there is no absolute union of *Atman* with *Paramatman* (Brahman); the practice of Yoga enables the embodied soul to realize its true nature, viz., eternal Self (Atman) that is same as Brahman. The false identification of the soul with the body, mind and intellect is loosely described as the feeling of separation from *Paramatman* and also as the bondage of *jivatman*. As the result of the practice of Yoga the *jivatman* is enlightened and realizes its oneness with *Paramatman*. So in its technical sense, **Yoga refers to that enormous body of spiritual values, attitudes, precepts and techniques (that have been developed in India over several millennia), which purify the mind and heart of a human being and enable him to realize his true nature, the Divinity within**.

It is the modifications or thought-waves of the mind that stand in the way of *jivatman* recognizing its Divine nature. Therefore the great sage **Patanjali** of ancient India, who codified the science of Yoga in his *Yoga Sutras*, **defines Yoga as "the stoppage of the modifications of the mind"** (*Yoga Sutras*, I.2). The modifications of the mind can be restrained by the persistent practice of Yoga and dispassion (*Vairagya*) (*Y.S.*, I.12). (For details, *see* 2.3 Mental Modifications and their Control, below). When the modifications of the mind are restrained and the mind becomes pure and focused on the higher Self, the Yogi becomes established in his essential nature, i.e., he attains Self-realization or Liberation.

Lord Krishna in the *Bhagavad Gita*[1] says, "...**Evenness of mind is called Yoga**" (2.48). Evenness means a balanced state of mind in pairs of opposites such as praise and blame, pleasure and pain, success and failure, and so on. Essentially it is an attitude of looking dispassionately at life without being ruffled by its ups and downs. The balanced state is another aspect of the purity of the mind.

The Paths of Yoga

There are several paths to the realization of the Self, just as there are many paths to the summit of a mountain. The **primary paths** are ***Karma Yoga*** (the Path of Selfless Service), ***Bhakti Yoga*** (the Path of Devotion), ***Jnana Yoga*** (the Path of Knowledge) and ***Raja Yoga*** (the Path of Meditation). The word "Yoga" in a generic sense refers to all the different paths. The paths vary so as to suit the varying temperaments and capacities of human beings, but all the paths lead to the same goal of Self-realization. The different paths of Yoga are not mutually exclusive. They simply represent a difference of emphasis.

Karma Yoga

The Sanskrit word *karma* means action. So Karma Yoga means Yoga of Action. The tendency to perform action is innate in all beings, and none can remain inactive. Of those who are particularly of active temperament, what is required is not cessation from action but the proper control and direction of their activity. Work with attachment to its fruit binds the soul, but selfless work performed in a spirit of worship becomes a means for attaining freedom. Selfless action does not bind the doer, as he has no sense of doing, and has no attachment to the fruits of his action. Lord Krishna in the *Bhagavad Gita* says,

> "One who does actions, offering them to the Divine, and forsaking attachment is not tainted by evil, just as a lotus leaf is not tainted by water" (5.10)
>
> "Though ever engaged in actions, one, having taken refuge in Me (the Lord), attains to the eternal and immutable state of being by My grace" (18.56)

The work that is offered to God becomes worship and Yoga. Karma Yoga thus enables a person to live successfully and usefully in the world while remaining above it, unaffected by its fetters, as lotus leaves on the water.

Bhakti Yoga

This Yoga of Devotion is a path of love and devotion to God. The Bhakti Yoga is simple and suitable for all those of emotional nature. It is an intense longing and love for God. It enables the aspirant to constantly remember God. It purifies his emotions and elevates his mind to the awareness of the Reality. He surrenders all his thoughts, words and deeds to the Lord and adores Him with an unflinching devotion. By constant meditation of the Lord, the *bhakta* (devotee) imbibes the Divine attributes into his own being. The devotee yearns for Lord's vision, and for his merger with his beloved God. Worldly love does not satisfy this deep passion of the soul. The devotee's entire heart flows, as it were, in a continuous stream towards the Lord. The Lord is not a being dwelling somewhere in a far off heaven, but is seated in his own heart. In order to see Him and to feel His Divine presence he has only to look within. Lord Krishna in the *Bhagavad Gita* says,

"The Lord dwells in the hearts of all beings…Take refuge in Him with all thy being; by His Grace you shall attain supreme peace and the eternal abode (18 61,62).

By total devotion and self-surrender to God, one can attain God or the ultimate Reality.

Jnana Yoga

The word *Jnana* means knowledge. So *Jnana Yoga* is Yoga of Knowledge. This path involves intense discrimination (*viveka*). One with a contemplative mind tries to seek answer to the basic question: "What am I? Am I a body or mind? If not, what else?" He discriminates between the transient and the eternal. Such persistent discrimination and self-inquiry ultimately lead to Enlightenment or the Knowledge of Self. Ramana Maharishi, the Indian sage of the twentieth century, was a great Jnana YogI Speaking of the followers of the path of knowledge, Sri Krishna in the *Bhagavad Gita* observes,

> "Those who have their intellect absorbed in That (Brahman), whose Self is That, with That as their supreme goal attain to the

Highest salvation, their impurities being cleansed by Knowledge" (5.17).
"Relative existence, even here in this world, is conquered by those, whose minds rest in evenness. As Brahman is spotless and equal, they are established in It" (5.19).

Raja Yoga

This "royal"(*Raja*) path is highly scientific. It is Yoga of Meditation. The saint Patanjali codified it in his *Yoga Sutras*. Raja Yoga encompasses the teachings of all the different paths; it concerns itself with three realms—the physical, the mental and the spiritual. By practicing Raja Yoga regularly, one learns to control his desires, emotions and thoughts and thus purifies his mind, and finally attains Samadhi, the state of Super-consciousness (*See* 3.10. Samadhi, below).

One with contemplative bent of mind and tremendous will power seeks to control the whole of nature, external and internal by practicing Raja Yoga. But irregular practice cannot be successful. The restless mind is to be brought under control gradually and steadily by persistent practice and dispassion. Sri Krishna in the *Bhagavad Gita* says,

> "With the intellect steadfast, with the mind fixed on the Self, let the Yogi attain quietude little by little, without thinking of anything else (6.25).
> Due to whatever cause the restless, unsteady mind wanders away, let him restrain it from wandering, and bring it back under the control of the Self (6.26).

Verily supreme bliss comes to the Yogi, whose mind is perfectly tranquil, whose passions are quieted, who is free from impurities, and who has become one with Brahman" (6.27). (For details, *See* 3.2 Raja Yoga: An Outline, below)

Complementary nature of the Paths

The **different paths** of Yoga are **not mutually exclusive**. They are **complementary** to one another. An aspirant cannot really follow any one of these four paths exclusively; he may only lay greater emphasis upon one or the other. For instance, Meditation is to be practiced as a discipline, no matter which path the aspirant may follow. Besides, an aspirant must cultivate discrimination and gain spiritual knowledge to practice even Bhakti Yoga and Karma Yoga. And even an aspirant who practices Bhakti Yoga or Jnana Yoga or Raja Yoga has to be active and perform actions without expecting their fruits or dedicate them to the Lord. Above all, every aspirant, irrespective of the path chosen by him, must have devotion to God or to the chosen Ideal. Thus in reality there is a combination of Yogas in every aspirant's life, because perfection is attained only by developing all aspects of one's personality—physical, mental, moral and emotional.

One-sided development is fraught with some dangers. Work sometimes becomes aimless and leads to mere restlessness. Yoga of Meditation at times degenerates into physical mortification and pursuit of psychic powers. Devotion often deteriorates into meaningless sentimentalism. Knowledge may also lapse into dry intellectualism and morbid inactivity. Hence there is a great need for combining the different paths in order to safeguard against the dangers of following one path exclusively. Over emphasis upon any one path leads to lop-sided development and disturbs the balance of life. Let work be combined with meditation, and knowledge be tempered by devotion. Let a Yogi be equally established in all the paths so as to bring about a harmonious development of all his latent spiritual powers. This all-sided development of the personality is the healthiest ideal for any spiritual aspirant to pursue. That is why, the teachings of the *Bhagavad Gita* and Sri Ramakrishna[2] emphasize synthesizing all the paths of Yoga in one's life. In the Integral Yoga Institutes established by Sri Swami Satchidananda, the apostle of peace, in the United States of America and other countries, this total approach is rightly called "Integral Yoga." **Integral Yoga** is a combination of Hatha Yoga, Pranayama, Raja Yoga, Bhakti Yoga, Jnana Yoga and Karma Yoga. It gives Yoga an affirmative and dynamic form. Its goal is complete self-integration.

It will lead to a balanced development, perfection and permanent peace and joy.

Action, devotion, wisdom and peace are inter-dependent and inter-related. Devotion, in its spiritual essence, is an attribute of wisdom. It flows from the vision of the inter-dependence of all life and the oneness of all existence. Similarly, knowledge is inseparable from devotion and action. Knowledge is a comprehensive awareness of the nature of existence. It is the vision of oneness of all things and beings. It awakens the Jnani to the timeless dimension of being. In the same way, action is not a mere means to self-purification. It is participation in the creative adventure of life. It is a total and joyful self-giving at the altar of the welfare of all beings.

Yoga and Religion

Yoga is not a special form of religion. Every religion in its essence is a kind of Yoga in so far as it helps human beings to realize God. Religions are different paths to the same goal. But unfortunately this is not understood by all people. Some become fanatics and indulge in religious persecutions. The sectarian creeds and ritualistic observations erect barriers among human beings and create divisions in the name of God, but Yoga stands for spiritual oneness of all beings and it is a universal spiritual pursuit beyond all religions. Everyone can practice Yoga, whatever may be his religious faith. The sacred scriptures of most religions contain references to Yogic methods. The mystics of all religions of the world have practiced Yoga in one form or other.

The Universality of Yoga

Yoga is universal in its approach. Everyone, irrespective of colour, class, greed, race, nationality, faith, etc. can practice Yoga. It is a means for achieving ethical, moral and spiritual perfection, which is our original nature. It dispels ignorance, develops universal outlook and love in the minds of the practitioners. Yoga enables them to perceive **unity in the midst of diversity**. The names and forms, thoughts, emotions, outlook, etc. vary but the Atman or Self within all beings is the same. The *Isa Upanishad*[3] declares,

"He, who sees all beings in own Self, and his own Self in all beings, loses all fear" (I.6).

Lord Krishna in the *Bhagavad Gita* says,

"He who sees Me (Brahman) everywhere and sees everything in Me, never becomes separated from Me, nor I become separated from Him" (6.30).

Since ages the saints of India and the mystics of other countries practiced Yoga and realized the ultimate Reality. The methods and techniques of Yoga have been tested and experimented by thousands of Yogis over centuries. Yoga enables a Yogi to go beyond the mind and experience the Self or Spirit, which is his real nature.

Though the practice of Yoga originated in India, Yoga is not confined to India. It has spread out to various other countries. A school of Meditation was established by Indian monks in Egypt around the third century B.C. Later the doctrines and practices of Yoga were carried to China and thence to Japan by Buddhist missionaries. The mystics of all major faiths such as Buddhism, Jainism, Judaism, Christianity and Islam have attained a direct insight into the nature of Reality. Thanks to Sri Ramakrishna Vedanta Centers established by Swami Vivekananda in Western Countries, and Yoga Centers established by great Indian Yogis of the last century in various parts of the world, Yoga is taught and practiced today by people belonging to various faiths all over the world, and it is bringing about a worldwide spiritual renaissance today.

Yoga and Asceticism (*Tapas*)

In India asceticism came into existence as an independent tradition alongside Yoga. It is a Yoga like practice. The Sanskrit word *Tapas* is derived from the verbal root *tap,* meaning, "to burn." The word *Tapas* came to be applied to the rigorous spiritual discipline of austerities. The ascetic gets "burnt out" through austerities. *Tapas* is a means to know Brahman. In the *Taittiriya Upanishad*, the teacher states to a student who wants to learn about Brahman,

"Through *Tapas* seek to know Brahman. Brahman is *Tapas*." (III.2.1)

The *Yajnavalkya Smrti* says,

"Of all the means employed to achieve objectives by disciplined means, *Tapas* alone is the most efficient one…*Tapas* means the stilling of the senses and the mind. The concentration of energies of the mind and the sense organs is supreme *Tapas*; that is verily superior to all disciplines."

"Tapas" is generally pursued through the observance of celibacy (*brahmacharya*) and the control of the senses. Celibacy is not suppression of sexual energy. It is a sublimation of spiritual energy, which generates psychophysical effulgence, radiance and vitality. It also generates numinous energy that charges the entire body.

Tapas yields vision of God, transcendental Consciousness and psychic powers. The parallel development of *Tapas* and Yoga is documented in the two ancient great epics of India—the *Ramayana* and the *Mahabharata*. In *Ramayana* we come across the stories of powerful demons like Ravana and his associates who witnessed their downfall by abusing their psychic powers acquired through severe *Tapas*. The *Mahabharata* narrates many stories of such celebrated ascetic sages as Vyasa, Vishvamitra, Vashishta, Bharadvaja and others.

Yoga spiritualized the early tradition of *Tapas* by emphasizing Self-realization over the acquisition of psychic powers. At the same time, it adapted many of the practices of *Tapas*. Celibacy is one of the five restraints (*Yamas*), which form the first step or limb of Raja Yoga. The sage Patanjali who codified the tradition of Yoga in his *Yoga Sutras* states that a Yogi who is grounded in celibacy gains vigour (*virya*) (II.38). He also mentions *Tapas* as one of the five observances (*Niyama*), the second step in Raja Yoga, and declares that through asceticism the body and its senses are perfected (II.43). The tradition of *Tapas* has existed alongside Yoga over centuries and continues to exist even today (*See* 3.4 Yama and Niyama, below).

Yoga and Renunciation (*Sanyasa*)

Though many aspects of *Tapas* have been integrated into the practice of Yoga, Yoga is closer in spirit to another tradition called Renunciation (*sanyasa*) of worldly life. Renunciation is a way of life. It is also an ancient tradition in India.

There are four stages in life (***Asramas***), namely, *Brahmacharya* or the period of studentship, *Grihastha* or the stage of householder, *Vanaprashta* or the stage of the forest dweller or hermit, and *Sanyasa* or the life of a renunciate. These four stages regulate the human life. Renunciation is of two forms—formal renunciation and mental. Formal renunciation means giving up ordinary worldly life. The renunciate leaves everything behind—spouse, children, property, work, worldly aspirations and concern for the future, and becomes a monk, seeking Self-realization. Mental renunciation means giving-up all attachments and egoism as well. The formal renunciation without dispassion and non-attachment is a mockery. Such a renunciate is a hypocrite, says Sri Krishna in the *Bhagavad Gita* (3.6). The formal renunciation with dispassion for worldly things enables the renunciate to concentrate on spiritual practices with total dedication.

An aspirant who seeks Self-realization through Yoga is expected to control the senses and give up desires for worldly objects and pleasures, though not worldly life. So mental renunciation is absolutely necessary for Yoga practice. Purity of heart is a must for attaining Self-realization. The control of senses and dispassion is a means for purification of heart. "Blessed are those pure in heart, for they shall see God," says the *Holy Bible.*[4]

Questions

1. What is Yoga? Is it a science? Justify your answer in terms of what you know about a science.
2. How does Patanjali define Yoga? Compare his definition with that of the *Bhagavad Gita*.

3. What are the various paths of Yoga? Briefly describe each of them. Are they mutually exclusive?
4. What is Integral Yoga? How is it important?
5. Is there any relationship between Yoga and Religion?
6. Is Yoga Universal? How?
7. What is *Tapas*? What is its connection with Yoga?
8. What is Renunciation? How is it important for Yoga?

1.2: Yoga Tradition in India

Pre-classical Period

Origin of Yoga

Yoga is as ancient as human existence. Lord Siva[5] Himself is a great Yogi.The earliest beginnings of Yoga have been lost in the obscurity of ancient Indian pre-history. The *Bhagavad Gita*, which was supposed to have been composed during the third or second millennium BC. speaks of Yoga as being archaic. It declares,

> "This (Yoga), handed down thus in regular succession, the royal sages knew. This Yoga, by a long lapse of time, has been lost here, O Arjuna" (4.2)
>
> "That same ancient Yoga has been today taught to thee by Me, for thou art My devotee and friend. It is the supreme secret" (4.3)

The above verses indicate the antiquity of Yoga and how it was handed down in regular succession from teacher to student. As time progressed much was added, much was changed by experimentation and in due course it was developed into a scientific psycho-spiritual system in ancient India. The ancient sages experimented with various approaches to Self-knowledge and discovered some definite methods of Yoga. The *Svetasvatara Upanishad* says,

> ". . . By meditating on the Lord, by uniting with Him, by reflecting on His being more and more, one becomes totally free from the illusion (*maya*) of the world " (I.10).

The knowledge so experienced was passed on from teacher to students in the oral tradition. This knowledge was realized not only by sages in depths of forests or caves but also by royal sages (ruling monarchs) who had to command armies, to sit on thrones and look after the welfare of citizens. In various *Upanishads* this is indicated. For example, Shvetaketu, the son of a sage was taught by Pravahana Jaivali, the king of the Panchalas. Similarly sages sent their sons to the royal sage, King Janaka for learning. The *Bhagavad Gita*, which is the essence of the *Upanishads*, was taught to Arjuna by Lord Krishna on a battlefield.

Indus Civilization

The earliest archaeological evidence for the practice of Yoga in ancient India is afforded by the magnificent Indus Civilization of India, discovered in the early 1920s during the Archaeological Survey conducted by Sir John Marshall. This civilization was one of the great civilizations of antiquity. It flourished in the fourth or third millennium BC. The survey found a number of soapstone seals that show horned deities seated in the posture of Yoga of Meditation. These finds have demonstrated that Yoga was dated to a period earlier to the *Vedic* period.

Yoga in *Vedas*

The *Vedas*, ancient Sanskrit scriptures of India, consists of truths revealed by God to the ancient Seers (*Rishis*) of India. They were handed down orally direct from teacher to disciple. Eventually they were committed to writing. The *Vedas* are divided into *Rig, Sama, Yajur* and *Atharva*. Each of them consists of two main divisions—the work portion and the knowledge portion. The work portion contains *mantras* (hymns) in praise of personal gods and the methods of using the *mantras* in sacrificial rites. The knowledge part of the *Vedas* is known as *Upanishads* or Vedanta (the end of *Vedas*). For details, see the next subheading in this Chapter.

The *Vedic* hymns are expressions of deep spirituality. The prototype of Yoga of the *Rishis* contains many elements characteristic of later Yoga: Concentration, watchfulness, austerities, regulation of breath, visionary experience and Samadhi.

Of the 1028 hymns of the *Rig Veda*, five are of special relevance to Yoga. The first is *nasadiya-sukta,* the hymn of creation (X.129). This hymn of cosmogony foreshadows the metaphysical speculation of the Samkhya Philosophy, which is closely allied to Yoga.

The second hymn (X.72) states how the universe came into existence. The third is the *Purusha-sukta* (X.90). In the first verse, the primeval *Purusha* is said to have pervaded the entire creation and extended beyond it.

The fourth hymn (I.164) is a collection of profound mystical riddles. The sixth verse refers to the nature of Self or Atman. Two famous verses 20-22 speak of two birds that occupy the same tree; one is said to eat its fruits, while the other merely looks on. The tree represents the body. The on-looking bird is the supreme Self beyond the realm of Nature, and the other bird is the individual embodied soul, which enjoys the fruits of actions and enmeshed in worldly life. In verse 46 of the same hymn contains the of-quoted statement, "*Ekam sat, vipraha bahudha vadanti.*" It means "Truth is one; the sages call it by various names."

The fifth hymn (X.136) known as *keshi-sukta* refers to the practice of *Pranayama* (regulation of breath).

The *Atharva Veda* contains references to immortal Self, three *gunas*, *Pranayama*, initiation, spiritual practice, non-dualism, etc. It also speaks of a community called *Vratyas* who laid the foundation of Yoga and Samkhya.

The *Yajur Veda* says that Yogis who have attained the state of immortality abide in the Lord. It also says that one acquires the knowledge of the past, future and present by the practice of Yoga.

Yoga in *Upanishads*

The word "*Upanishad*" is derived from "*upa*" (near), "*ni*" (down) and "*sad*" (to sit), i.e., sitting down near. Groups of students sit near the teacher to learn from him the sacred doctrine. In the forest hermitages the seers communicated their "revealed" wisdom to fit students sitting before them.

The *Upanishads* contain the essence of the *Vedas*. They are rich in profound mystical thought. "In the whole world," says Schopenhauer, a renowned philosopher of the West, "there is no study so beneficial and so elevating as that of the *Upanishads*. They are products of the highest wisdom. They are destined sooner or later to become the faith of the people."

The *Upanishads* contain the inner direct, intuitive and mystical experiences of the ancient seers of India, revealed by God. The truths revealed to seers are verified not only by logical reason but also by personal experiences. According to tradition there are one hundred and eight *Upanishads*. The principal *Upanishads* are said to be ten. Shankaracharya[6] commented on eleven, viz., *Isa, Kena, Katha, Prasna, Mundaka, Mandukya, Taittiriya, Aitareya, Chandogya, Brahadaranyaka* and *Svetasvatara*. These represent the *Vedanta* Philosophy in its pure original form, and are the earliest philosophical compositions of the world.

According to them, Brahman is the one single Reality. It is the source of all existing things. It is known through Atman, our inner Self.

The *Upanishads* contain important conceptions peculiar to Yoga and its ally Samkhya Philosophy. The sages of *Upanishads* have adopted the practice of meditation as the chief means of obtaining transcendental knowledge. They emphasized the need for renunciation, intense contemplation, discrimination between the real and unreal and burning desire for Self-realization for experiencing the Self.

The *Upanishads* speak of not only the glory of the Self and the means of realizing It, but they also teach other things including the cycle of births and deaths governed by the moral Law of Karma, and the esoteric means by which the world of change and the endless rebirth can be transcended.

Brahadaranyaka Upanishad: In this *Upanishad*, Yajnavalkya describes *Pranayama* (III.4.1). When being asked how the Self is to be conceived, he states,

". . . You cannot see the Seer of seeing. You cannot hear the Hearer of hearing. You cannot understand the Understander of understanding. He is your Self, which is in all things. Everything else perishes" (III.4.2).

This passage epitomizes the essence of the *Upanishads'* mystic teachings that were passed on from Self-realized sages to disciples by word of mouth. The transcendental source of the universe is identical with the transcendental core of the human beings. That is, Brahman equals Atman. That supreme Reality cannot be described. It must simply be realized. Upon realization, the Self will be found to be infinite, eternal, free and blissful.

Isavasya Upanishad: It prescribes two paths: the path of renunciation or Jnana Yoga for renunciates, and the path of Action or Karma Yoga for others.

Kena Upanishad: It deals with the nature of Brahman and the Knowledge of Self. It also explains how one can realize the Brahman by transcending the mind and senses. In its fourth part, it explains the method of meditation and its benefits.

". . . Seeing It (Brahman) in all beings, the wise become immortal" (II.5).

Katha Upanishad: This *Upanishad* explicitly deals with Yoga of Self. It explains fully the way to attain Self-realization in the Section 3 of Chapter I. Only when a Yogi goes deep inside, withdrawing from senses, sense-objects, mind, intellect and the unmanifested *Prakrti*, he can know his own Self. Yoga means the state of inner tranquility. When the five senses together with the mind cease from their normal activities and the intellect becomes still, that is the highest state. This is Yoga, the steady control of senses. (II.3.10, 11). The teachings of the *Katha Upanishad* represent an important milestone in the tradition of Yoga. Its beautiful poetry expresses some of the fundamental ideas underlying Yoga.

The ***Prasna Upanishad*** lays great stress on the sacred syllable OM, the *Pranava*.

The ***Mundaka Upanishad*** teaches the highest knowledge of Brahman. An individual soul attains liberation through the Knowledge of Brahman. Just as the rivers, when flowing into the ocean, become one and the same with it, so also one who knows Brahman becomes Brahman himself. (III.2.8, 9).

The ***Mandukya Upanishad*** contains the essence of the wisdom of the *Upanishads*. It gives an exposition of the principle of AUM and represents Jnana Yoga in its highest form. It represents the entire non-dualistic tradition of the *Upanishads*.

The ***Taittiriya Upanishad*** deals with the Knowledge of the Supreme Self (Brahman).

> "He who is here in a human being, and He who is there in the sun, He is the One. He who knows this, on departing from this world, reaches the Self consisting of food, reaches the Self consisting of life-force, reaches the Self consisting of mind, reaches the Self consisting of understanding, reaches the Self consisting of bliss" (II.8.1).

This verse indicates the Vedantic doctrine of five sheaths (*pancha kosha*): food sheath (*annamaya kosha*), prana sheath (*pranamaya kosha*), mental body (*manomaya kosha*), intellectual sheath (*vijnanamaya kosha*), and bliss sheath (*anandamaya kosha*). These five sheaths cover the Atman. One who transcends all the five sheaths realizes Atman.

The ***Aitareya Upanishad*** states the creation of the world. The whole universe is the manifestation of Brahman, and the individual Self is identical with the Supreme Self; and the goal of life lies in the realization of the identity of the individual Self with the Supreme Self. The result of this realization is immortality.

Renunciation is necessary to practice the Knowledge of Brahman. A renunciate can devote the whole time in study and meditation. The practice of meditation cannot be pursued with intense vigor and devotion without relinquishing the concern of life.

The ***Chandogya Upanishad*** contains elaborate mystical speculations about the sacred syllable OM in Sections 1 and 2 of Chapter I, Section 23 of Chapter II.In his commentary on this *Upanishad* Shankaracharya observes that OM is the most appropriate name for the Divine or Transcendental Reality.

Svetaketu, a boy, wanted to know the Supreme Knowledge. His father explained this to him with the examples of a seed and salt placed in water. From the very essence in the seed, which we cannot see, comes the vast banyan tree. Similarly, salt dissolved in water cannot be seen. In the same way we cannot see the Cosmic Spirit. That is Reality. That is Truth. Thou art That.

The *Chandogya Upanishad* answers the fundamental question "What am I?" You are not the finite body-mind complex, but the infinite Atman that dwells within your heart.

The ***Brahadaranyaka Upanishad*** is recognized as the most important of the *Upanishads*. It consists of three Sections. The first Section expounds the identity of the individual Self with the Universal Self. The Second Section provides a philosophical justification for this exposition. The third Section deals with the methods of spiritual practices to be followed: hearing the teaching (*sravana*), logical reflection (*manana*) and contemplative meditation (*nididhyasana*).

In this *Upanishad,* the origin of the universe is said to spring from the one being, the Absolute Brahman or the Supreme Spirit. The realization of this One Being is the goal of life. It says,

> " . . . So verily the Supreme Being, infinite and limitless, consists of nothing but pure Consciousness" (II.4.12).

The Universal Self is not a thing that could be pointed to in the finite world. A steady practice of discrimination between the real and the unreal awakens in the Yogi the will to renounce everything in the world of change. Discrimination and renunciation ultimately lead to the realization of the Universal Self.

The ***Svetasvatara Upanishad*** deals with Karma Yoga and Jnana Yoga. It is theistic in character. Personal God and Devotion to Him, the elements of theism are prominent in this *Upanishad.* Terms that were used by the later Samkhya Philosophy occur in this Upanishad, but *Prakrti* (Nature) is not an independent entity as in

Samkhya. This *Upanishad* teaches the unity of the souls and the world in the One Supreme Being. It describes Liberation in the following verse:

> "By knowing God, there is falling of all fetters; when the sufferings are destroyed, there is cessation of birth and death. By meditating on Him, there is the third state; on the dissolution of the body, Self, being alone, all its desires are fulfilled, and it becomes One without a second (I.11)

This verse describes the different sides and stages of Liberation. Negatively, it is freedom from the cycle of births and deaths; positively it is Oneness with God so long as there is the manifested world, and it is Oneness with Brahman when the manifested world ceases to exist.

The *Svetasvatara* recommends meditation on OM, the *pranava*. It describes the meditative process as a kind of churning by which the inner fire is kindled, leading to the revelation of the Self's splendor. It suggests the pre-requirements of right environmental conditions, correct posture, regulation of *prana* and concentration (II.8-10)

When the mind is stilled, all kinds of preliminary visions—fog, smoke, sun, wind, fire, fireflies, lightning, moon, may appear (II.11). Among the first signs of progress of Yoga are lightness, health, steadiness, clearness of complexion, pleasantness of voice, agreeable odor and scanty excretions (II.13).

In the verse II.6 the synthesis of Karma Yoga, Jnana Yoga and Raja Yoga is suggested. Karma Yoga will purify the mind; Raja Yoga will steady the oscillating mind; and Jnana Yoga will remove the veil of ignorance. The next verse treats Bhakti Yoga. Without *bhakti* (devotion) you cannot attain the Grace of God, which is essential for attaining the Knowledge of Brahman. When you attain Brahman, you are not bound by your former actions. The fire of knowledge burns the seeds of karmas.

The ***Kausitaki Upanishad*** contains a detailed exposition of the doctrine of rebirth and a description of the path to the "World of the

Absolute." It also contains a long discourse on the life force as being identical with the Absolute. One verse reads as:

> "Life is *prana, prana* is life. . . For, indeed, through *prana*, one obtains even in this world Immortality" (III.2)

The ***Maitrayaniya Upanishad*** describes the mystery of the Great Self and the individual soul. The individual soul is affected by the good and bad effects of actions, by the pairs of opposites like pleasure and pain. It is subject to the cycle of births and deaths. But it can realize the supreme Self through study, austerity, recitation, meditation on OM and the pursuit of one's own duty. It also expounds the six-fold Yoga: *Pranayama,* withdrawal of the mind from senses, meditation, concentration, contemplative inquiry and absorption (*Samadhi*). It mentions the central channel, *Sushumna Nadi,* along which *prana* must rise upward from the base of the spine to the crown of the head. This process is accomplished by joining the breath, the mind, the senses, and the sacred syllable OM.

The ***Maitrayaniya*** hints at practices that suggest a further advance in the development of Yoga, and thus prepared the ground for Patanjali's classical formulations.

Yoga in Epics

The two great ancient Epics of India are the *Ramayana* and the *Mahabharata*. No single literary work has more influence on the lives of millions of people in India and Southeast Asia than the *Ramayana*, the story of Sri Rama.

Ramayana: The significance of the *Ramayana* for the students of Yoga lies in the moral values it promulgates vividly. It can be regarded as a consummate treatise on *Yama* and *Niyama*, the first and second limbs of Raja Yoga codified by the sage Patanjali in his *Yoga Sutras*. The *Ramayana* extols virtues like righteousness, non-hurting, love, dispassion, humility, truthfulness, charity, chastity, and other virtues and disciplines like austerities. It also serves as a great illustration for Karma Yoga, selfless service. Sri Rama's brother Lakshman, and His devotee Hanuman are illustrious Karma Yogis.

The *Ramayana* also illustrates how Yogis who misuse psychic powers acquired through intense *Tapas* (austerities) destroy themselves. Ravana and his associates had acquired enormous psychic powers through intense Yogic practices, but they are demons and applied their powers for mean selfish purposes and for causing untold miseries to innocent people including celestials. They were killed in the war with Sri Rama, the embodiment of righteousness and Divine qualities.

Yoga Vaishta: This is a continuation of *Ramayana*. This philosophical work of about 32,000 verses is the earliest work on Yoga and Vedanta. It contains the teachings of the sage Vasishta to the young prince Rama. The path of Yoga outlined in this work is essentially Jnana Yoga. The Buddhi Yoga taught in the *Bhagavad Gita* is similar to it. The genuine Yogi, according to Vasishta, is free from the pull and push of the passionate attraction and aversion. Such a Yogi looks at a lump of gold and a pile of rubbish with the same equanimity. It is the human mind that creates bondage and liberation. The mind is compared to a mad man with thousand hands, who constantly beats himself. The mind is galvanized by the vibration of the life-force (*prana*) circulating in the body. Regulation of the *prana*, concentration and meditation are the means for quieting the mind and transcending the compelling force of desires.
Vasishta's Yoga comprises seven steps:

(1) eagerness to know the truth,
(2) deepening of study and contact with holy people,
(3) refinement of thinking,
(4) dispassion or non-attachment,
(5) quiescence of the mind,
(6) recognizing the illusion of the world and the reality of the Self, and
(7) Liberation or abiding in pure Consciousness (Self).

The Yogi who realizes the Self is liberated while living. Enlightenment is transcendence of ego. Vasishta says,

"Just as there is butter in every kind of milk, the Supreme abides in the bodies of all things."

"The perverted intellect, who considers himself as the body, is verily confined in it, but when he knows himself to be identical with the taintless Self, he is liberated."

"With boundless patience, courage and service, carry on your meditation and self-study, and worship God in holy Yoga."

Mahabharata: The second epic *Mahabharata* is a magnificent and invaluable treasure house of mythology, religion, philosophy, ethics, customs and stories of kings and sages. It comprises about 100,000 stanzas. The nucleus of this Epic is the strife between righteous Pandavas and their cousins wicked Kauravas. Ultimately truth prevails.

The *Mahabharata* is an encyclopedia of Hindu righteousness. It contains noble moral teachings, useful lessons of all kind, many beautiful stories, episodes, discussions, parables and dialogues that set forth the principles of righteous living, morals and metaphysics. Yoga and Samkhya are the highlights of the philosophy of this Epic. The most significant materials on Yoga are found in the famous *Bhagavad Gita* and the *Moksha Dharma* Section, which form part of the sixth and the twelfth Books of the *Mahabharata.*

Bhagavad Gita: The *Bhagavad Gita* or Song Celestial is part of the epic *Mahabharata*. It is a dialogue between Lord Krishna, God-incarnate, and his disciple Arjuna, one of the Pandava brothers. This immortal conversation is the climax of the story of *Mahabharata*. The *Bhagavad Gita* contains the cream of the *Upanishads*. It is a universal gospel. It is of great importance to the Yoga tradition. It teaches the Yoga of Knowledge of Self, Yoga of Meditation, Yoga of Knowledge, Yoga of Action, Yoga of Self-control and Yoga of Devotion.

Aldus Huxley calls this ancient work as "the most systematic statement of the perennial philosophy." The *Gita* is the first successful attempt in human cultural history to work out a complete philosophy of life, reconciling the secular and sacred, work and worship. Even though the *Gita* is an ancient scripture composed several thousand years back, it is relevant to all times and to all people.

The central message of the *Gita* is performance of one's duty with dispassion or non-attachment and balanced in success and failure (2.48). This means harmonization of the worldly life with spiritual life.

In order to attain enlightenment, Lord Krishna declares, one need not forsake one's responsibilities. Renunciation of action is good in itself, but better still is renunciation in action. This is the celebrated ideal of inaction in action (*Niskarmya Karma*), which is the basis of Karma Yoga. Life in the world and spiritual life are not in principle inimical to each other; they can go on simultaneously. Such is the essence of an integrated life.

Sri Krishna taught the *Gita* to Arjuna as his charioteer in the battlefield of Kurukshetra. It symbolizes the eternal battle raging within the mind of everyone, a battle between good and bad, virtue and vice. Arjuna represents the individual soul; the charioteer represents the Lord, the Divinity within our heart; the chariot represents the body; the horses represent the five senses; and egoism, mental impressions, desires, likes and dislikes, lust, greed, anger, jealousy, pride, hypocrisy etc. are our enemies. In this battle the message of the *Gita* will ever remain as an unfailing guide to the individual souls.

The essence of the teachings of the *Gita* relating to Yoga is briefly presented below.

In the **Second Discourse** Sri Krishna teaches Jnana Yoga and describes the immortality of the Soul (Self) in verses 2.11-38. The body is impermanent and perishable, but the Soul or Spirit within it is immortal. It always exists, even after the death of the physical body (2.12). The Self pervades all objects like ether (2.17). It is existence Absolute. Just as a person casts off worn-out clothes and puts on new ones, so also the embodied soul casts off worn-out bodies and enters new ones (2.22). The eternal Self is indivisible. It is extremely subtle. Weapons cannot cut It, fire cannot burn It, water cannot wet It, and wind cannot dry It (2.23, 24). The body of any creature may be destroyed, but the Self that dwells within it cannot be killed (2. 30).

Sri Krishna also describes the characteristics of a sage of steady wisdom and the means of attaining that steady knowledge of the Self in verses 2.55–72.

In the verses 2.47-51 of the **Second Discourse** and in the **Third Discourse** Sri Krishna teaches Karma Yoga (Yoga of Action). Your right is to work only but not to its fruits or results (2.47). If you have desire for the fruits of your actions, you have to take birth again and again to enjoy them, both good and bad effects of your good and bad actions. Actions done with the expectation of rewards bring bondage. Therefore perform actions dispassionately without any motive, you will attain purity of heart, and finally, Self-realization (3.19). In reality all actions are performed by the three *Gunas* (*Sattva, Rajas* and *Tamas*) of Nature. Through ignorance only, one thinks that he is the doer of all actions.

> "All actions are wrought in all cases by the *Gunas* of Nature only. He, whose mind is deluded by egoism thinks, "I am the doer" (3.27)..

Sri Krishna teaches the Yoga of Wisdom in the **Fourth** and **Fifth Discourses**. If an aspirant attains wisdom through discrimination and spiritual practice, the fire of wisdom burns all his mental impressions and impurities and purifies him. So one who attains Knowledge of the Self and takes refuge in God becomes absorbed in Him.

> "Freed from attachment, fear, and anger, absorbed in Me, taking refuge in Me, purified by the fire of knowledge, many have attained to My Being" (4.10).

It is the idea of "I am the doer" that binds you to the worldliness. If you identify with actionless Self and observe the actions of you body-mind system as a silent witness, feeling that Nature does everything and that you are the non-doer, then the actions done are no actions at all. This is **inaction in action**. On the other hand, you may sit quiet, not doing anything, but if you think that you are the doer, then you act though you sit quiet. Your mind acts. This is **action in inaction.**

> "He, who sees inaction in action and action in inaction, is wise among men; he is a Yogi . . ." (4.18).

A liberated sage may do service to others as worship of God, without any attachment, then his actions are no actions at all (4.23).

Just as monks who have renounced the world attain Liberation through the path of Jnana Yoga, so also Karma Yogis engaged in selfless service as offerings to the Lord attain Liberation (5.5), because their actions do not bind them. Their minds are purified, thereby they acquire knowledge of the Self (5.10).

The sages, who behold the Self only everywhere, look with equal vision on a learned Brahmin, on a cow, and even on a dog and an outcaste (5.18). Such sages are established in Brahman. They discover happiness in the Self (5.21). They attain Liberation while living (5.24).

Sri Krishna teaches the Yoga of Meditation in the **Sixth Discourse**. He describes the Meditation posture, Concentration, control of the senses, the need for keeping the mind balanced and for observing moderation in eating, working, recreation and sleeping. He also describes the benefits of Yoga (6.27, 28). The Yogi, whose mind is serene and free from passion and attachments enjoys supreme bliss. He sees the Self, abiding in all beings and all beings in the Self. He, who sees Brahman (God) everywhere and everything in Him, dwells in Him (6.30). Sri Krishna also points out the restlessness of the mind and the need for controlling it by practice and dispassion (6.34, 35).

Who is the best Yogi? Sri Krishna says, "Among all the Yogis, one, who with full faith and merging his inner self in Me, worships Me, is the best Yogi" (6.47).

In the **Ninth Discourse** He describes the Knowledge of the Self and Self-realization. In the **Tenth Discourse** the Divine nature and glories are described. In the **Eleventh Discourse** the Vision of the Cosmic Form is described.

> "He, who does all actions for Me, who looks upon Me as the Supreme, who is devoted to Me, who is free from attachment, and who bears no enmity towards any creature, come to Me" (11.55)

This verse contains the essence of the philosophy of the *Gita*. When you perform actions as offering unto the Lord you will attain purity of your heart and eventually realize God.

In the **Twelfth Discourse** Sri Krishna teaches the *Bhakti Yoga* (Yoga of Devotion). The devotees who fix their minds on God as the Supreme Lord and worship Him, ever steadfast and endowed with supreme faith are the best in Yoga (12.2). Those, who worship the Supreme Lord having restrained all the senses, even-minded and working for the welfare of all beings, also realize Him (12.3, 4). Those, who worship the Supreme Lord, with unswerving exclusive devotion, offering all actions to Him, regarding Him as the supreme goal, and meditating on Him with single-minded Yoga, attain immortality (12.6, 7).

Those, who are unable to fix their minds steadily on the Lord, can seek to reach Him either by the Yoga of constant practice of remembrance or by doing actions for God's sake, or by renouncing the fruits all actions and practicing self-control (12.9-11). The Knowledge of the Self, gained from the study of the scriptures, is better than ignorantly pursuing practices (12.12).

The qualities and qualifications of devotees dear to the Lord are described in the verses 12.13–20. These are: loving all creatures with compassion, no hatred towards anyone, free from attachment and egoism, balanced state of mind, contentment, steady meditation, firm conviction, self-control, total devotion and dedication to the Lord, non-hurting others, forbearance, free from mental modifications of joy, envy, fear and anxiety, free from dependence, indifference to the body, senses and objects of senses, renunciation of all undertakings or personal initiatives and good and evil, sameness to foe and friend and pairs of opposites, contentment with bear means of sustenance, faith, and regarding the lord as the supreme goal.

In the **Thirteenth Discourse** Sri Krishna teaches the difference between *Prakrti* and Spirit (Self) in detail. In the **Fourteenth Discourse** He describes the nature of three *gunas* and how one can go beyond the *gunas* and attain Self-realization. In the **Fifteenth Discourse** the Lord teaches the real nature of the Self. In the **Sixteenth Discourse** the divine and the demoniacal qualities are described.

The **Seventeenth Discourse** deals with the three kinds of people who are endowed with the three kinds of faith. Each of them follows a path in accordance with his inherent nature—*sattvic, rajasic* or *tamasic*. The *sattvic* persons worship the gods; the *rajasic* worship the spirits that guard wealth and demons with illusive powers; and the *tamasic* worship the ghosts. A person's food, sacrifice, austerity and charity are also determined according to his *guna.*

In the **Eighteenth Discourse** the teaching of the whole of the *Gita* is summed up beautifully. The distinction between renunciation and sacrifice, and the Yoga of Liberation are also described. The verse 18. 65 sums up the essence of the entire *Bhagavad Gita:*

> "Fix your mind on Me, worship Me, dedicate every action to Me, and surrender yourself to Me. Then you will certainly come to Me. I promise you this because you are my devotee" (18.65).

This verse reflects the integration of all kinds of Yoga. Each phrase indicates one of the paths of Yoga Integrate all kinds of Yoga in your life and take refuge in the Lord, He will take care of you and take you beyond sufferings and sorrow, i.e., you will attain eternal Liberation, peace and bliss.

Moksha-Dharma*:* Next to the *Bhagavad Gita*, the most significant materials on Yoga in the *Mahabharata* are found in the *Moksha Dharma* Section. This Section comprises Chapters 168–353 of the Twelfth Book. *Moksha-Dharma* gives important details about the Epic forms of (or Pre-classical) Samkhya Philosophy and Yoga. Both traditions have their specific theoretical frameworks and they are distinctively theistic. The Epic Yoga is based on solid moral foundations. It instills such virtues as truthfulness, humility, non-possessiveness, harmlessness, forgiveness and compassion. Lust, anger, greed and fear are listed as the Yogi's greatest enemies. Doubt, discontent and psychic powers are deemed to be severe obstacles on the spiritual path. Withdrawal of senses from the external world, concentration and meditation are considered to be the primary means of Yoga.

Yoga in Ethical Scriptures (*Dharma Sastras*)

Elements of pre-classical Yoga[7] are also found in the Ethical Scriptures known as *Dharma Sastras*. The highest spiritual life can blossom only when it is founded on morality. Therefore there are many references to Yoga in the manuals on ethics and law, which regard Liberation as the highest possible goal. The *Manu Smriti* is an important work of pre-Christian era on morality and law. It says, "Desires are not satisfied by their fulfillment; the more one tries to satisfy them, the more they grew." They multiply and lead to restlessness, fear, anxiety and constant worry. Therefore this scripture says, "Freedom from desire is total happiness." It recommends Pranayama, Concentration and Meditation for combating such undesirable emotions as anger, avarice and jealousy. According to Manu, there are ten values of *dharma* to be followed by all people. They are: fortitude or forbearance, forgiveness, control over organs of action, non-stealing, purity, mastery over the senses, discrimination, knowledge, truthfulness and absence of anger. Patanjali's prescription of *Yamas* and *Niyamas* are similar to these values. Similar statements are found in the *Baudayana Dharma Sutras*, an important juridical work of the same period.

There is a quotation of a verse in *Apastamba Dharma Sutras* of fourth century B.C., which states that all taints of character can be eliminated through the practice of Yoga.

The *Yajnavalkya Smriti* describes the entire Yogic Process—from assuming the right posture, to withdrawing senses from external objects, to performing Pranayama, Concentration and Meditation.

What is clear from these Ethical Scriptures is that Yoga was already an integral part of India's cultural and moral life in pre-Christian era.

Classical Period

Patanjali's *Yoga Sutras*

The sage Patanjali codified the teachings on Yoga into *Yoga Sutras*. This work represents the culmination of the long development of the system of Yoga over millions of years. The growth of this system in ancient India is fascinating as a phenomenon. It mirrors the germination of numerous ideas right from the pre-Vedic period.

What Patanjali did was to bring together various principles and practices of Yoga prevalent at the time into one coherent system. He gave it a structure and method and made it as an independent philosophical system. He has put the theoretical formulations into an experiential mode by laying down practical methods, which everyone could practice and realize the ultimate goal of Liberation.

What was Patanjali's period? What was he? Hardly anything is known about him. The author of *Yoga Sutras* and the author of *Mahabhasya* (grammar) are both called Patanjal I. Some scholars believe that the authors of both works are the one and the same person. Others deny this. Scholars' estimates of the date of the *Yoga Sutras* vary widely from the fourth century BC to the third century AD.

The *Yoga Sutras* contains 195 *Sutras* and it is divided into four *padas* (parts): *Samadhi Pada* (Samadhi), *Sadana Pada* (Practice), *Vibhuti Pada* (powers) and *Kaivalya Pada* (Liberation).

The word "*Sutras*" means threads connecting a work together. They are short forms with a few essential words each. Often there is no complete sentence structure. The form of *sutras* was used in ancient days when there were no books, for the sake of easy memorization and study by the students. Therefore the *sutras* need extensive explanations in order to make their meanings clear to the students

In the **first *pada*** the theory of Yoga (I.2-16) and the kinds of Samadhi (I.17-22; I.41-51) are described in detail; the obstacles to Yoga and the various means to overcome them are also explained (I.30-40). *Sutras* I.23-28 define the nature of *Isvara* (God). Patanjali admits the existence of God, which Samkhya Philosophy did not accept. Many alternate methods for practice of Yoga are described in *sutras* I.32-39. This shows that Patanjali brought together the many Yogic beliefs prevalent at that time.

In the **second *pada*** the eightfold (limbs) path of Raja Yoga is enumerated and described meticulously up to the fifth stage (limb) of *Pratyahara* (II.29-55). The first five steps are external means to Yoga, while the next three (Concentration, Meditation and Samadhi), collectively called "*Samyama*" are the internal means and are described in the third *pada* (III.1-7).

The **third *pada*** describes the last three steps (limbs) of Yoga (III.1-15) and the supernatural powers that come naturally to one who practices *Samyama* (III.16-55). The acquisition of these powers is not the real goal of Yoga. So Patanjali warns the Yogis not to be attracted and distracted by the psychic powers, as they can obstruct them from reaching the ultimate goal of Liberation (*Kaivalya*). Thus the emphasis on Liberation as the final aim of the Hindu Philosophy is reiterated.

The **fourth *pada*** deals with the final goal of Liberation. It also describes the world of birth and death, the nature of karma, the difference between *Purusha* (Self) and the *Sattva* intellect and the objects of experience (IV.25-34). Some concepts that were dealt with in the first and second *padas* are also repeated in the fourth *pada*. Works of this kind are bound to have a certain amount of overlapping of ideas and concepts.

Thus Patanjali has admirably brought together under his scheme of eight-limbed Yoga the many Yogic beliefs and practices that were probably important during his time. The system richly deserves the title of Raja Yoga.

Commentaries: It would be very difficult to understand the *sutras* without the help of a commentary. Vyasa's *"Bhashya"* is the earliest commentary. The next available commentary is Shankara Bhagavadpada's *Vivarana* on the Vyasa's *Bhashya.* The subsequent commentaries include: Vacaspati Mishra's *Tattvavaisaradi* (nineth century AD), Bhoja's *Bhojavrtti* (eleventh century), Vijnabhiksu's *Yogavarttika* (sixteenth century), Ramananda Sarasvati's *Maniprabha* (sixteenth century).

In recent times a few English translations with commentaries on the *Yoga Sutras* have come out. These include: Extempore comments made by Swami Vivekananda during his classes on Patanjali held in U.S.A during 1890s (these were written down by his students, and included in his book on *Raja Yoga*); Swami Prabhavananda and Christopher Isherwood's commentary entitled "*How to Know God, the Yoga Aphorisms of Patanjali*" (1969); I.K. Taimni's *The Science of Yoga* (1967); P. N. Mukerji's translation of *Swami Hariharananda Aranya's Commentary on Yoga Sutras of Patanjali* (1963); Sri Swami Satchidananda's translation and commentary "*The Yoga Sutras of Patanlali*" (1978) and T.S. Rukmani's commentary on *Yoga Sutras of Patanjali* (2001). These are listed under "References."

Srimad Bhagavatam

This magnificent scripture depicts the heroic life of Lord Krishna and the stories of sages, devotees and kings. No other scripture, except the *Bhagavad Gita*, has enjoyed such widespread popularity through centuries in India. The devotional literature *Bhagavatam* is pre-eminent for its philosophical depth, devotional exuberance and literary beauty.

Yoga is mentioned in many passages of the *Bhagavatam*. It reconciles, like the *Bhagavad Gita*, Yoga of Devotion, Raja Yoga, Yoga of Knowledge and Yoga of Action. "It is fried in the butter of Knowledge," says Sri Ramakrishna, "and steeped in the honey of Love." The *Bhagavatam* accepts Patanjali's eight limbs of Yoga, but rejects his dualistic philosophy.

Of particular interest for the Yoga student is the Book XI known as the *Uddhava Gita*. In it Lord Krishna expounds the Yoga of Devotion to His disciple Uddhava.

Perhaps the most extraordinary teaching of the *Bhagavatam* is the Yoga of Hatred. According to this Yoga, a person who thoroughly hates the Divine can achieve God-realization as readily as one who deeply loves the Lord. Sage Narada, frequent speaker for the *Bhagavatam* teaching, expresses it thus:

> "All emotions are grounded in the erroneous conception of 'I' and 'mine.' The Absolute, the Universal Self, has neither 'I' sense nor emotions" (7.1.23).
>
> "Hence one should unite (with God) through friendship or enmity, fearlessness or fear, attachment or love. (The Divine) sees no distinction whatsoever" (7.1.25).

Kamsa reached God through fear, and Shishupala through hatred. The idea that hatred can turn out to be a pathway to God, though shocking apparently, it is a logical consequence of the ancient esoteric doctrine that we become whatever we meditate upon.

Post-classical Period

Most of the Yoga teachings that emerged after Patanjali's period did not adopt his dualistic metaphysics of *Purusha* (Self) and *Prakrti* (Nature). They adopted the Vedantic non-dualism, i.e., God or Brahman alone is the only Reality.

The post-classical literature of Yoga is most diversified and richer in content than the pre-classical literature. It includes: *Puranas*, *Saiva Agamas* of the Saivism, one of the main branches of Hinduism, *Vishnu Samhitas* of Vaishnavism, another main branch of Hinduism, and *Tantras* of the Shakti worshippers.

Puranas (Mythology)

These are ancient **narratives** that help us to understand the theme of the *Vedas* **through** historical and metaphorical **stories**. The theme and purpose of the *Vedas*, the Epics and the *Puranas* is the same, i.e., to reveal the *Paramatman*, Absolute Reality. The *Puranas* reveal the Truth through the discussion of the origin of five elements and all other things created out of the basic elements and through moral values. They teach **moral values** through stories.

The *Puranas* contain, among other things, brief treatment of Yoga and stories about aspirants and masters of Yoga. The *Markandeya Purana* speaks in detail about the qualities of an aspirant of Yoga and also about the environmental conditions necessary for success in the practice of Yoga.

The *Linga Purana* describes Patanjali's eight-stage Yoga and the obstacles to be overcome.

The *Agni Purana* contains extensive information about *mantra* recitation, *Mudras* (hand gestures), *Yantras* (mystic diagrams), *Pranayama*, and Patanjali's eight-stage Yoga. In the *Garuda Purana* an important place is assigned to the eight-stage Yoga and also to *Bhakti Yoga* and *Tantric Yoga.*

The voluminous *Shiva Purana* deals with five types of Yoga including *Mantra Yoga* and *Maha Yoga* (Great Yoga), which is contemplation of Lord Shiva without any restricting conditions.

The *Puranas* thus contain records of various Yogic practices. Some of them follow Patanjali's model. However, what distinguishes them from Patanjali's tradition is that they propose a single ultimate principle, the Self or God. Many of the *Puranas* are available in English translations. The fund of myths and legends preserved in the *Puranas* is a perennial source of inspiration to the Yoga students.

The *Agamas*

These are **theological** treatises and **practical** manuals of Divine **worship**. They include the *Tantras, Mantras* and *Yantras*. These treatises explain the external worship of God. They also give elaborate details about ontology and cosmology, Liberation, Devotion, Meditation, philosophy of Mantras, mystic diagrams, etc. There are *Agamas* for each of the three sections of Hinduism, viz., Saivism, Vaishnavism and Saktism. The *Saiva Agamas* glorify God as Siva and have given rise to an important school of philosophy known as *Saiva Siddhanta*. The *Vaishnava Agamas* glorify God as Vishnu. The *Sakta Agamas* or *Tantras* glorify God as Sakti or the Mother of the Universe.

Darsanas

The *Darsanas* are **Schools of Philosophy** based on the *Vedas*. They appeal to the intellect, while the Epics, *Puranas* and *Agamas* appeal to the heart. The word *darsana* means "the vision of truth."

The chief Hindu philosophical systems are Nyaya, Vaisesika, Samkhya, Yoga, Mimamsa and Vedanta. Each of these *Darsanas* holds a holistic approach towards life, aiming at a direct experience of the ultimate Reality or Truth. Yoga is closely related to the Samkhya Philosophy and Vedanta. (For detail, *see* **1.3 Indian Philosophy: An Overview**, below, and **1.4 Samkhya Philosophy and Patanjali's Yoga Philosophy**, below)

Shankaracharya

Shankaracharya (eighth century AD) is one of the greatest sages in the history of Yoga. He made great contribution to the Philosophy of Vedanta and Yoga. His major writings were his commentaries on eleven *Upanishads*, on the *Bhagavad Gita* and on the *Vedanta Sutras*. He also wrote original mystical poems such as *Viveka Chudamani* (Crest Jewel of Wisdom), *Aparokeshanubhuti* (Direct Experience of Reality), *Atma Bodha* (Knowledge of the Self) and *Vakyavritti* (Explanation of the Text). As a great sage and a Yogi, he handled difficult metaphysical questions with relative ease and simplicity. The world is fortunate that almost all his writings are now available in English translation.

The Southern Saivism

The Southern Saivism favors a qualified monism that, in practice, is dualistic with Lord Siva and the devotees. This is epitomized in the devotional songs of such great saints as Appar, Sundarar, Manickvasagar and Thirugnana Sambandar. The Sekizhar's *Periapuranam*, which tells the life stories of sixty-three *Nayanamars* (Saiva Saints), is a great literature on Bhakti Yoga.

The famous ***Tirumantiram*** of the great Sage Tirumular is an extraordinary scripture on Yoga, devotion, *Tantric* practices and gnosis (*jnana*). It is a unique scripture of 3,000 verses spread over nine *Tantras*. Tantra three describes the procedure of eight-limbed Yoga and special types of Yoga such as *Kecari Yoga* (the Supreme state of Yogic Bliss) and *Chantra Yoga* (Yoga for achieving immortality). Tirumular discusses Yoga more fully than PatanjalI.Tantra Four deals with *Mantras* and *Yantras*. Tantra

Seven among other things deal with Kundalini Yoga. Tantra Nine presents a picture of the Divine Vision. (See Tirumular, *Tirumantiram* (with English Translation and Notes), Madras: Sri Ramakrishna M ath, 1991).

The Northern Saivism

The Northern Saivism leans towards a non-dualist interpretation of the Reality. At the beginning of the ninth century the Kashmiri adept Vasagupta discovered the *Shiva Sutras*, a digest of the earlier Agama teachings. This is a treatise on Yoga. The commentaries on the *Shiva Sutras* contain invaluable materials on *Pranayama*.

Samhitas

These are the religious works of the Vaishnavas. They correspond to the *Saiva Agamas* and the *Tantras* of the Sakti worshippers. The emphasis of the teachings of the *Samhitas* is on **ritual worship** and **moral way of life**. The *Vishnu Samhita* introduces a six-fold Yoga. But many other *Samhitas* recommend Patanjali's eight-stage Yoga. The devotional poems of Alvars (Vaishnava saints) sparkle with passionate love and devotion for God. The Alvars were steeped in intense devotion to Lord Krishna, a full incarnation of the Lord Vishnu. Their "rapturous passions are like a whirlpool that eddies through the very eternity of the individual soul." The Alvars in their ecstatic delight visualize God everywhere.

The *Gita Govinda*, Sanskrit poems of Jeyadeva of Bengal (North-eastern part of India), is a profound allegory of the love between the Personal God and the human Self.

The works of Vaihnava preceptors include *Yoga Rahasya* (Secret Doctrine of Yoga) by Nathamuni, *Siddhi Traya* (Triad of Perfection) by Yamuna and the works of the great Ramanuja, expounding the philosophy of qualified monism. Yogic teachings also played a role in the works of other great Vaishnava preceptors such as Madhva (1238–1317), Nimbarka (mid-twelfth century) and the ecstatic Krishna Caitanya (1486–1533). These teachers are great exponents of Bhakti Yoga.

Questions

1. Yoga is considered to be of Divine origin. Substantiate this statement.
2. Is there any Archaeological evidence for the practice of Yoga in the ancient period?
3. What are *Vedas*? What aspects of Yoga are emphasized in the *Vedas*?
4. What are *Upanishads*? How do they describe the Self or Atman?
5. What is the relevance of each of the important *Upanishads* to Yoga?
6. What is Brahman? How can one realize Brahman?
7. What is the relevance of the *Ramayana* and *Yoga Vasishta* to Yoga?
8. What is the importance of *Mahabharata* to Yoga?
9. Write a small essay on the teachings of *Bhagavad Gita* about Yoga.
10 What is the relevance of *Dharma Sastras* and *Puranas* to Yoga?
11. What is the contribution of Patanjali to the science of Yoga?
12. What is the relevance of *Srimad Bhagavatam* and *Agamas* and *Samhitas* to Yoga?
13. Discuss the contributions of Tirumular and Shankaracharya to Yoga.

1.3: Indian Philosophy: An Overview

Introduction

Philosophy, in its widest sense, means "love of knowledge." It is termed in Indian literature as *Darsana,* which means "vision of truth." Indian philosophy, unlike modern Western philosophy, which is primarily an intellectual pursuit, is based on knowledge of truth. It tries to seek answers to fundamental questions of life: What is the purpose of life? What is the nature of the world in which man lives? and so on.

Indian philosophy is not a mere metaphysical speculation. It is a practical **guide to** the daily **life**. Its insights were directly experienced by ancient sages of India. Its theories and ideas are not limited to any particular era, culture or group. They are universal, and are meant for one and all.

Indian philosophy is much more comprehensive than Western philosophy. It is an **integration of several disciplines** like metaphysics, epistemology, logic, axiology, aesthetics, ethics, sociology, psychology and physiology. In the West different aspects of every problem are studied by separate disciplines. But Indian philosopher studies each problem from all possible approaches. This tendency denotes the synthetic or total outlook of Indian philosophy. It holds a **holistic approach** toward life, aiming at a direct experience of the ultimate Reality and Truth. It is a body of knowledge that frees one from fear and from the primitive instinct of self-preservation. The ultimate Truth can be experienced not by a chosen few only, but under right conditions, by all human beings.

Modern man knows much about the external world, but knows

nothing about his real and internal nature. Prompted by external clamors and sophistication he seeks peace and happiness from external objects and sensual enjoyments, and suffers from emotional problems and several psychosomatic diseases.

This is due to a lack of a philosophy of life. Life, without a guiding philosophy, is like a castle built on sand. Indian philosophy provides an effective **philosophy of life**, and opens the door to eternal peace and joy.

The systems of Indian Philosophy

The Schools of Indian Philosophy are divided into two **broad** groups: **Orthodox** and **Heterodox**. To the first group belong the six main philosophical systems, namely, Nyaya, Vaisesika, Samkhya, Yoga, Mimamsa and Vedanta. These are regarded as orthodox because they accept the authority of the *Vedas*.

The three main heterodox systems are Jainism, Buddhism and Carvaka. These do not believe in the authority of the *Vedas*. However they do believe in the authority of their own spiritual leaders and founders.

Though the different Schools hold different views, each School took care to learn from the views of all others, and did not come into any conclusion before considering thoroughly what others had to say and how their points could be met. This catholic spirit of treating rival positions with consideration was more than rewarded by the thoroughness and perfection that most of the Indian Schools attained.

There is a gradual **progression from** concern with the scientific methods for investigation of the material aspects of the universe to concern with the duality of the universe in terms of matter and consciousness—and finally, in the Vedanta, to a concern with the essential nature of Consciousness itself.

The Nyaya, the Vaisesika, the Samkhya and the Carvaka have based their theories on ordinary human experience and reasoning. But on matters such as God, Liberation, etc., ordinary experience is not dependable. Philosophy must depend for these on the experience of saints and seers who have a direct realization of such things.

The authority or the testimony of reliable persons and scriptures thus forms the basis of philosophy. The Mimamsa and the Vedanta Schools follow this method. Even the Buddhism and the Jainism depend mostly on the teachings of their prophets.

The *Upanishads* are the greatest source of Indian philosophical thought and all the systems of Indian philosophy are found in their potential form in these works. The philosophical expositions in the *Vedas* , however, are expressed in the form of aphorisms, which are brief, terse and abstruse. The theories take an explicit form only in the *Upanishads*.

The Carvaka System

The word *Carvaka* means a " materialist." The Carvaka system believes in pure materialism. Its origin can be traced as back as the *Vedic* period. The Carvaka is mentioned in the Epics and in the Dialogues of Buddha. The main work on the system, the *Brahaspati Sutra* (600 B.C), is not available, and the doctrines of materialism have been reconstructed from the statements found in polemical and other works. The *Tattvopaplavasimha* (seventh century AD) is the only extant treatise on the Carvaka system.

The Carvaka's doctrine is called *lokayata*, as it holds that only this world exists and there is nothing beyond it; there is no future life. The postulates of religion, God, freedom and immortality are illusions. The highest end of life is enjoyment of pleasure in this life.

The Carvaka holds that perception is the only valid source of knowledge. Perception reveals to us only material world, composed of the four elements: air, fire, water and earth. There is no evidence to show the existence of a soul in man. The elements of matter give rise to the living body . When a man dies nothing is left of him to enjoy or suffer the consequences of his actions hereafter.

Jainism

The origin of Jainism lies far back in the prehistoric times. The long line of teachers who handed down this faith consists of twenty-four Tirthankaras, liberated saints. The last Tirthankara was Varthamana (also known as Mahavira), a contemporary of Gautama Buddha.

Jainism admits perception, inference and testimony as sources of valid knowledge. On the basis of these sources it forms its view of the universe . The physical world consists of four elements: matter, space, time and causes of motion and rest. Souls exist in all living bodies.

The actions (*karmas*) lead to the bondage of the soul. Three things are necessary for the removal of the bondage, namely, perfect faith in the teachings of the teachers of Jainism, correct knowledge of the teachings, and right conduct.

The right conduct consists in the practice of abstinence from all injury to life, falsehood, stealing, sensuality and attachment to sense objects. When the *karmas* that bind the soul are removed, the soul attains its natural perfection.

Jainism does not believe in God, but the Thirthankaras are adored as ideals of life. The system does not accept the authority of the *Vedas*.

In course of time, the followers of Jainism were divided into two Sects known as Svetambaras and Digambaras. The teachings of Tirthankaras are accepted by both the Sects. But the Digambaras are more rigorous and puritanical, while the Svetambaras are more accommodating to the common frailties of human beings.

The major works of Jainism are: Umasvati Acarya's *Tattvarthadhigama Sutra*, Mallisena's *Syadvadamanjari* and Siddhasena Divakara's *Sanmati tarka.*

Buddhism

Buddhism is based on the teachings of Gautama Buddha (sixth Century BC). As a prince he was awakened to the awareness of human sufferings, and he wanted to find the means to overcome the sufferings. He spent several years in study, penance and meditation to discover the origin of human sufferings and the means to overcome them. At last he received enlightenment and thus discovered the Four Noble Truths: 1. Suffering exists; 2. There is a cause of suffering; 3. Suffering can be eradicated; 4. There is a means for the eradication of suffering.

Buddha developed an Eightfold Path as a means for the

eradication of suffering. The Eightfold Path consists of right views, right determination, right speech, right conduct, right livelihood, right endeavor, right mindfulness and right concentration. These eight steps remove ignorance and desires, enlighten the mind and bring about perfect equanimity and peace. Thus liberation from misery and rebirth is attained. This state of perfection is called *nirvana.*

After Buddha's death his followers modified his original teachings and formed two major divisions: Hinayana and Mahayana. Hinayana holds more rigidly to the original teachings of Buddha. It accepts Buddha as an enlightened teacher, and *nirvana* as the goal of life.

Mahayana has integrated the teachings from other philosophical systems and holds the attainment of perfect wisdom as the object of *nirvana.* Mahayana consists of two philosophical Schools, namely Suniyavada or Madhyamika School and Vijnanavada or Yogacara School. According to the Suniyavada School, the world is unreal;mental and non-mental phenomena are all illusory. This view is known as Nihilism. The Vijnanavada School holds that external objects are unreal. What appears as external is really an idea of the mind. But mind must be admitted to be real. This view is called Subjective Idealism.

Hinayana consists of two philosophical Schools: Sautrantika School and Vaibhasika School. Sautrantika holds that both mental and non-mental are real. External objects can be inferred to exist outside the mind. This view is called Representationism. Vaibhasika School agrees that both internal and external objects are real. But it differs from the Sautrantika School regarding the way the external objects are known. External objects, according the Vaibhasikas, are directly perceived and not inferred from their ideas or representations in the mind. For if no external objects were perceived corresponding to any idea, it would not be possible to infer its existence from any idea. This view is called Direct Realism.

The main works of Buddhism are the *Dhammapada, Samyutta-nikaya, The Dialogues of the Buddha and Visuddhi-magga.*

The Nyaya System

This was founded by the sage Gautama. It is a philosophy based mainly on logical grounds. The word *nyaya* in popular usage means " right" or "just" and Nyaya becomes the science of right or just reasoning. It is, in a wider sense, the science of correct knowledge. The main works of Nyaya system are: the *Nyaya Sutras* of Gautama (third century BC), the commentaries on the *Sutras* and Udayana Acarya's *Kusumanjali* (tenth century AD). The Nyaya philosophy admits four sources of valid knowledge, namely perception (*pratyaksa),* inference (*anumana*), comparison (*upamana*) and testimony (*sabda*). All other systems of Indian philosophy use the Nyaya system of logic for philosophical reasoning and debate.

The existence of God is proved by the Nyaya system by several arguments. God is the ultimate cause of creation, maintenance and dissolution of the world. He created the world out of eternal atoms, space, time, ether, minds and souls for the good of all living beings. The individual souls act for good or bad ends, and thereby bring happiness or misery on themselves. But under the loving care and guidance of the Divine Being, all individuals can sooner or later attain Self-knowledge and liberation.

The Vaisesiska System

This system was founded by the sage Kanada (third century BC). It is allied to the Nyaya system. It has the same end in view, namely, the liberation of the individual soul. It is mainly a system of physics and metaphysics. It brings all objects of knowledge under seven categories: substance (*dravya*), quality (*guna*), action (*karma*), generality (*samanya*), uniqueness (*visesa*), inherence (*samavaya*) and non-existence (*abhava*). This school is called Vaisesiska because it considers *visesa* (uniqueness) as an aspect of reality and studies it as a separate category. Under the category of substance, the Vaisesika deals with the physics and chemistry of the body and universe. The theory of atomic structure was established by this School. Its practical teaching emphasizes *dharma*, the code of conduct that leads man to the worldly welfare and to the highest goal of life.

Each soul experiences the consequence of its own deeds due to ignorance. Liberation is attained through right knowledge of reality.

The main works of the system are: Kanada's the *Vaisesika Sutras* (third century BC) and Prasastapada's *Padarthadharma samgraha* (fourth century AD).

The Samkhya System

This is a system of dualistic realism, attributed to the sage Kapila. The Samkhya is considered to be the oldest of all philosophical systems. It admits two ultimate realities: conscious *Purusa* and unconscious *Prakrti*. *Purusa* is ever pure and ever free, but it becomes the subject of pain and pleasure when it forgets its true nature and identifies itself with *Prakrti*. *Prakrti* is the material cause of the universe and is composed of three *Gunas*—*Sattva*, *Rajas* and *Tamas,* which correspond to light, activity and inertia respectively. (*See* 2.5 Three Gunas, *below)*

The disturbance of the equilibrium of *gunas* produces the material world, including the mind. The process of the world's manifestation passes through twenty-three successive states. The Samkhya philosophy also explains the dynamics of the body and the nature of the mind (See 1.4. Samkhya Philosophy and Patanjali's Yoga Philosophy, *below*).

Although the *Purusa* or Self is in itself free and immortal, due to ignorance it identifies itself with the body, the senses and the mind, and thus experiences pain and suffering. Once it realizes its true nature through spiritual practice, it attains freedom from suffering. The Samkhya does not admit the existence of God.

The main works of this system are: Isvarakrsna's the *Samkhya Karika* (third century AD) and the commentaries on it.

The Yoga System

Yoga was practiced even in pre-Vedic ages. But it was not formally systematized until it was codified by the sage Patanjali in his *Yoga Sutras* during third or second century BC. The Yoga philosophy accepts the epistemology and the metaphysics of the Samkhya with its twenty-five principles, but, unlike the Samkhya, it admits the existence of God. What the Samkhya calls *mahat* is called *citta* (mind-stuff) in the *Yoga* system.

The *Yoga* system studies all aspects of human personality and teaches one how to control the modifications of the mind through the observance of ethical discipline, practice of meditation, detachment and surrender to God.

It prescribes an eight-step holistic system of practice. These are: *Yama* (restraint), *Niyama* (observance), *Asana* (physical posture), *Pranayama* (breathing exercises), *Pratyahara* (withdrawal of senses), *Dharana* (Concentration), *Dhyana* ((Meditation) and Samadhi (absorption). The individual Self ultimately finds the pure Consciousness within by practicing Yoga.

The major works are: Patanjali's Yoga *Sutras*, Vyasa's *Yoga Bhasya* (fourth century AD) and Vacaspati Misra's Tattva-*Vaisaradi* (850 AD) (*See* 1.4 Samkhya Philosophy and Patanjali's Yoga Philosophy, below)

The Mimamsa System

The Mimamsa system was founded by Jaimini. His *Mimamsa Sutra* outlines the details of this system. The primary objective of the Mimamsa is to defend and justify *Vedic* ritualism. The authority of the *Vedas* is the basis of ritualism, and the Mimamsa formulates the theory that the *Vedas* are not the works of any person and are, therefore, free from errors. The *Vedas* are eternal and self-existing. For establishing the validity of the *Vedas*, the Mimamsa discusses the theory of knowledge. The validity of *Vedic* knowledge is self-evident. The Mimamsa is foremost in the analysis of sound and mantra.

In this system, the whole life is considered to be a grand ritual. The rituals enjoined by the *Vedas* should be performed not with the hope of any reward but just because they are so enjoined. The disinterested performance of the rites gradually destroys the *karmas* and brings about liberation.

The Mimamsa admits soul as immortal; but consciousness is not intrinsic to it; it arises in the soul when it is embodied . The Mimamsa does not believe in the Supreme Soul or God.

Knowledge is derived from perception, inference, comparison, testimony and postulation.

The Mimamsa believes in the reality of the physical world on the strength of perception. But it does not accept the cycle of creation and dissolution. The world exists always. Its objects are formed out of matter in accordance with the *karmas* of the souls. The Law of *Karma* is an autonomous natural and moral law that rules the world. Liberation for the Mimamsa is life in heaven, and not the state of ultimate release found in most other systems.

Jaimini admits the reality of the *Vedic* deities, but does not argue for the existence of a supreme God. Some later Mimamsakas admit the reality of God.

The Vedanta System

The Vedanta is the most comprehensive of the Indian systems of philosophy. This system arises out of the *Upanishads,* which mark the culmination of the *Vedic* speculation and are fittingly called the *Vedanta* or the end of the *Vedas*. The Vedanta was taught and practiced by the sages of the *Vedas* and *Upanishads* and was handed down through a long line of sages. Badarayana Vyasa codified these teachings in the *Brahma Sutra*. In seventh century AD. Shankara wrote commentaries on the *Brahma Sutra* and important *Upanishads*.

There are three main Schools of Vedanta systems: non-dualism, qualified non-dualism and dualism.

Shankara's Non-dualism (*Advaita Vedanta*): Of all the systems, the *Advaita Vedanta*, as interpreted by Shankaracharya, has exerted the greatest influence on Indian life. According to this School of Vedanta, the Supreme Brahman is One without a second. It is the only Reality. It pervades the whole universe and yet remains beyond it. The world originates from this eternal Soul, rests in it and dissolves in it. We perceive the many objects on account of our ignorance, which conceals the real Brahman from us and makes it appear as many objects. Brahman embodies within itself the power of Maya, which is the basis of mind and matter.

The main teachings of Advaita Vedanta are: Self-realization is the goal of life. The individual Self is essentially Brahman. The soul realizes its true nature when ignorance is destroyed by knowledge attained through intense practice of discrimination, dispassion and meditation. The realization of the Truth " I am Brahman" is perfect wisdom or Liberation from bondage.

Ramanuja's Qualified Non-dualism: The teachings of Vedanta are interpreted and developed by Ramanuja (eleventh century) as *Visistadvaita* or Qualified Monism. According to him, God is the only Reality; yet within God there exist material objects and conscious souls as parts of the world. An individual soul becomes attached to the world through ignorance. When it is enlightened, it is liberated from bondage. Salvation, according to Ramanuja, is not the disappearance of the Self, but its release from bondage. The Self cannot be dissolved into God. The released Self has a permanent intuition of God.

Madhva's Dualism: Madhva (1197–1276) holds that God, selves and the world exist permanently, but the latter two are subordinate to God and dependent on Him. Brahman or God possesses all perfection and is identified with Vishnu. The Supreme God directs the world.

For Madhva, everything on earth is a living organism. The Self is not an absolute agent, since it is of limited power and is dependent on God. Though it is by nature blissful, it is subject to pain and suffering on account of its connection with a physical body due its past *karma*. The divine will is supreme. It sets men free or casts them into bondage. Salvation, for Madhva, is the perpetuation of the individual Self in the condition of Liberation, where the Self takes delight in adoration and worship of God.

The Development of the Systems of Indian Philosophy

In the West the different schools have come into existence successively. One School predominates till another comes in and replaces it. In India we find that the different Schools have flourished together and pursued parallel courses of growth. This is because in India philosophy was a part of life. As each system came into existence, it was adopted as a philosophy of life by a band of followers. They lived the philosophy and handed it down to succeeding generations of followers who were attracted to it through their lives and thoughts. The different systems of philosophy thus continued to exist through unbroken chains of successive followers for centuries.

Indian philosophy has had an extremely long and complex development and probably a longer history of continuous development than any other philosophical tradition. It is impossible to present an exact historical survey of this development, because many of the details of the chronological sequence of the writings are not available.

Nevertheless, Indian philosophy may be said to have five major periods of development.

1. The *Vedic* Period: This is the first stage of development of Indian philosophy. This period may be placed approximately at 3,000 BC. The Mantras of *Vedas* constitute the actual beginning of Indian philosophy. The *Rig Veda* presents various philosophical ideas of a general nature, and there are three regular currents of philosophical thought, which represent the Samkhya , the Vedanta and the Nyaya systems of later periods. In the *Rig Veda* are also traced the beginnings of the Tantric current with the Nada Brahman as the root cause of the universe.

The *Upanishads*, the last part of the *Vedas*, have established the tendency to spiritual monism which characterizes much of Indian Philosophy and where intuition rather than reason was first recognized as the true guide to the ultimate truth.

2. The Epic Period: The Epic period is dated approximately 2,500 BC. This period is characterized by the indirect presentation of philosophical doctrines through the medium of non-technical literature, especially the great Epics, the *Ramayana* and the *Mahabharata.* The *Ramayana* is an epic poem composed by Valmiki. It narrates the story of Rama of Ayodhya. The *Mahabharata* narrates the great battle between the Kauravas and the Pandavas. The most important section of the *Mahabharata* is the *Bhagavad Gita*, a philosophical dialogue between Krishna and Arjuna.

The Epic period witnessed the rise of early development of Buddhism, Jainism, Saivism and Vaisnavism. The *Bhagavad Gita* ranks as one of three most authoritative texts on Indian philosophical literature. The beginnings of the orthodox Schools of Indian philosophy also belong to this period.

It was also during this period that many of the *Dharma Sastras*, treatises on ethical and social philosophy were compiled. The *Dharma Sastras* are instructions in the sacred law. Of them the *Manu Smriti* (the laws of Manu) is the most famous work. It contains a philosophy of classes and stages in life. A *Smrti* is that which is remembered, while the *Vedas*, and the *Upanishads* are known as *Srutis* or revealed scriptures or authoritative texts.

3. The Period of Sutras: This period is dated approximately from third or second century BC. In this period, the systematic treatises of the various schools were written in the form of *sutras*, brief aphorisms.

The six Hindu systems presented in *sutra* form during this period are: (1) the Nyaya or logical realism; (2) the Vaisesika or realistic pluralism; (3) the Samkhya or evolutionary dualism; (4) the Yoga or disciplined meditation; (5) the Mimamsa or earlier interpretative investigations of the *Vedas* relating to rituals; (6) the Vedanta or later investigations of the *Vedas* relating to knowledge.

4. Scholastic Period: In this period commentaries were written upon the *Sutra*s in order to explain them. The *Sutra*s are very terse and their meaning is not clear without the proper guidance of a competent teacher. So competent teachers wrote commentaries on the *Sutra*s, which are called *bhasyas,* elaborate interpretations. There are *Bhasya*s on all the *Sutra* works of the different systems of philosophy.

This period has brought forth some of the greatest of the Indian philosophers. Among them are Shankara, Kumarila, Sridhara, Ramanuja, Madhva, Vacaspati, Udayana, Bhaskara, Jayanta, Vijnabhiksu and Raghunatha. These great thinkers were much more than commentators. They were creators of their own systems.

5. Modern Period: In the twentieth century, the original works and their commentaries were translated into English. The philosophies were analyzed, critically reviewed, and their histories were compiled in the forms of Encyclopedias, Source books, and

History of Indian Systems of philosophy. Great scholars of public and monastic life and professors of Indian and Western Universities have contributed to this monumental academic work. But for this work, the world would not have come to know about the eternal and universal ancient philosophy of India.

The Spirit of Indian Philosophy

Indian philosophy is extremely **complex**. Through the ages the Indian philosophical mind has probed deeply into many aspects of human experience and the external world. Not only themes of Indian philosophy but also the methods used and the conclusions reached in the pursuit of truth have certainly been far-reaching in their extent, variety and depth, as those of other philosophical traditions. The six basic systems and the many sub-systems of Hinduism, the four chief Schools of Buddhism, the two Schools of Jainism, and the materialism of the Carvaka, vouch safe the diversity and variety of views in Indian philosophy.

Nevertheless, in certain respects there is a distinct **spirit** of Indian philosophy. This is exemplified by certain distinctive **characteristics** of the Indian philosophy, which are described below.

1. Spiritual Approach: The chief mark of Indian philosophy in general is its concentration upon the **spiritual aspect of life**. Both in life and philosophy the spiritual motive is predominant in India. Except for the minor materialistic School of the Carvaka, philosophy in India conceives man as spiritual in nature, is interested primarily in his spiritual destiny, and relates him to a universe, which is also spiritual in essential character. Philosophy and religion are intimately related, because the motive in both of them is concerned with the spiritual way of life in the here-and-now and the eventual spiritual liberation of man in relation to universe.

2. Close relationship between philosophy and life: Another characteristic of Indian Philosophy is the intimate connection of philosophy with life. Philosophy never been considered a mere intellectual exercise. It did not originate merely in curiosity as it seems to have done in the West. It rather originated under the pressure of practical need to find a remedy for the ills of life. Every

Indian system seeks the ultimate remedy of freedom from pain and sorrow. This is not an arm-chair academic exercise, but a serious effort to discover the methodology for practical application in life. Each of the major systems of Indian philosophy searches for the Truth in order to solve the problem of man's distress in the world. In India philosophy is for life; it is to be lived. The goal of the Indian is not to just know the ultimate Truth , but to realize It, to experience It, to become one with It.

3. Introspective Approach to Reality: Indian philosophy is characterized by the introspective approach to reality. Philosophy is thought of as *Atmavidya*, knowledge of the Self. Physical science, though developed extensively in the Golden Age of Indian culture, was never considered to be the road to the ultimate Truth; Truth is to be sought and found within.

4. Idealism: This introspective interest is highly conducive to idealism, and consequently Indian philosophy is mostly idealistic in one form or another. The tendency has been in the direction of monistic idealism. Almost all systems of Indian philosophy believe that Reality is ultimately one and spiritual. However, Indian philosophy has not been oblivious to materialism; rather it has known it , has overcome it, and has accepted idealism as the only tenable view.

5. Intuition: Indian philosophy makes unquestioned and extensive use of reason and logic, but intuition is accepted as the only method through which the ultimate Truth can be known. Reason is insufficient. It can demonstrate the truth, but it cannot discover or reach the truth.

6. Direct Experience: The ordinary method of observation and rational inquiry is not suitable for studying one's own real nature and for realizing the ultimate Reality. After much experimentation, the ancient sages discovered the methods of meditation for internally reaching the highest level of Consciousness in which one could directly experience the Truth. Direct experience is, therefore, the foundation of all the systems of Indian philosophy.

7. Synthesis: In Indian philosophy, there is the overall synthetic tradition, which is essential to its spirit and method. All the six orthodox systems and their subsystems are in harmony with one another. They complement one another in the total vision, which is One. As contrasted with Western philosophy, with its analytic approach to reality, Indian philosophy is fundamentally synthetic. It studies man's entire being – body, mind, *prana* (vital energy) and Self, and his relationship with the external world. It adopts an integrated total approach to the study of man.

8. The Goal of Life: There is also a fundamental unity of perspective in the practical realm. The goal of life in all systems of philosophy in India is liberation, i.e., freedom from suffering and rebirth, and the attainment of infinite bliss.

9. Moral Order: the Law of *Karma*: All the Schools except Carvaka School of Materialism have firm faith in " eternal moral order" based on the Law of *Karma*. As per this law, all actions , good or bad, performed with a desire for the fruits thereof produce their proper effects in the life of the doer. His past actions determine his present life, and his future is shaped by his present actions. However, selfless action does not germinate. Man is the maker of his own destiny.

Almost all Schools regard the universe as the moral stage. The body, mind, senses and the environment that an individual gets are the endowments of Nature or God in accordance with the inviolable Law of *Karma*.

The Value of the Study of the Indian Philosophy

The study of Indian philosophy is important in the search for the Truth and India's contribution to the total picture of the Truth is unique indeed. From the study of Indian Philosophy, one can know the proper way of life, the ultimate goal of life; and he will be motivated to realize that goal and the oneness of all beings in the Atman by practicing Yoga.

The current appeal for "one world" or "the unity of the world" is too often thought of merely in the realm of politics. Political unity is impossible without philosophical understanding. "It is to philosophy that man must turn in his hope to bring the peoples of the world together in greater mutual understanding and in the intellectual and spiritual harmony without which a unified world will be impossible in any sphere, political or otherwise. The future of civilization depends upon the return of spiritual awareness to the hearts and minds of man. To this purpose, the contribution of Indian philosophy, with its age long spiritual emphasis, is inestimable and indispensable." (S. Radhakrishnan and C.A. Moore, 1973, p. xxxi)

Questions

1. How does the Indian Philosophy differ from the Western Philosophy?
2. How do you classify the systems of Indian philosophy?
3. What is the nature of gradual progression in the development of the Indian Philosophy
4. Give a brief account of the origin and philosophy of Jainism and Buddism.
5. What are the main aspects of Nyaya and Vaisesika systems?
6. Briefly describe the Samkhya system?
7. What is the objective of the Mimamsa system? What are the views of Mimamsa on soul, world and liberation?
8. How do you justify the name of the Vedanta system of philosophy ? What are its main teachings?
9. State the three main Schools of the Vedanta and discuss their main differences.
10. What is the contribution of the *Vedas* and *Upanishads* to the systems of Indian Philosophy?
11. What are the main characteristics of Indian Philosophy ?
12. Trace the developments of Indian Philosophy in different periods
13. How do you account for the coexistence of different systems of Indian Philosophy?
14. What is the importance of the study of Indian Philosophy ?

1.4: Samkhya Philosophy and Patanjali's Yoga Philosophy

SAMKHYA PHILOSOPHY

Introduction

The Samkhya Philosophy was expounded by the sage Kapila during the seventh century BC. The *Samkhya Pravacana Sutra*, which is now not available, is attributed to him. The earliest available text is the *Samkhya Karika* of Isvarakrsna of the third century AD The Samkhya is a philosophy of dualistic realism. It admits two ultimate realities:

> *Purusha (*Spirit or Self) and the unconscious *Prakrti* (primordial matter or Nature).

Purusha and *Prakrti*

Purusha is ever pure and ever free, but it becomes subject to pain and pleasure when it forgets its true nature and identifies itself with *Prakrti*. *Purusha* consists of multiple selves. *Prakrti* is both the material and the efficient cause of the world. It is composed of three *gunas* – *sativa* (light), *rajas* (activity) and *tamas* (lethargy). It is active and ever changing, but blind and unintelligent. In contact with the conscious and intelligent *Purusha*, *Prakrti* evolves the world. The Samkhya holds that the mere presence of *Purusha* is sufficient to move the *Prakrti* to act, although *Purusha* itself remains unmoved. Similarly it is due to the reflection of the conscious *Purusha* on the unconscious intellect (*manas*), the cognitive and other psychical functions are performed by the latter. But how the mere presence of *Purusha* can be the cause for changes in *Prakrti* is not clearly explained. Further the multiplicity of *Purusha* is proved with reference to the differences in the nature, activity, birth and death and sensory and motor endowments of different living beings. But all these differences pertain not to the Self as pure Consciousness but to the bodies associated with it. The pure Consciousness in all beings is the same. According to the Vedanta, there is only One Reality, namely, Atman and everything else—including *Prakrti*—is Its manifestation. The Vedanta thus carries the process of idealization to a further stage that alone can be the ultimate stage.

The Theory of Causation

The evolution of the universe is explained on the basis of a theory of causation known as *satkaryavada.* This means: the effect exists in its material cause in potential form prior to its production. There can be no production of a thing that is previously non-existent potentially in its material cause. The Samkhya system holds that the entire world, including our body, mind, senses and the intellect are the manifestations of *Prakrti.*

Prakrti and *Gunas*

Prakrti is a mere balanced combination of three *gunas* (forces): *sattva, rajas* and *tamas. Sattva* (goodness or purity) is potential consciousness; *rajas* (passion) is the source of activity, and *tamas* (dullness or inertia) is the source of resistance to activity. The three *gunas* are not the attributes of *Prakrti.* They are the constituent components of *Prakrti.* All things including human beings, as products of *Prakrti*, consist of the three *gunas* in different proportions. The three *gunas* are intertwined like three strands of a rope that binds the soul to the world. The varied interaction of the *gunas* accounts for the variety of the world. When there is equilibrium, there is no action. When there is a disturbance of the equilibrium, the process of evolution begins.

Prakrti is unconscious. Its evolution can take place only through the presence of conscious *Purusha*. The presence of *Purusha* excites the activity of *Prakrti*, and, thus, upsetting the equilibrium of *gunas* in *Prakrti*, passively starts the process of evolution.

Evolution of the Universe

At the advent of a new cycle of creation, *Prakrti* is roused from its slumber by the transcendental influence of *Purusha*. As a consequence of this, the equilibrium state of *gunas* is disturbed, and the *gunas* begin to vibrate. This primeval vibration releases a tremendous energy within *Prakrti*, manifesting the universe in twenty-four stages.

The first evolute of *Prakrti* in the process of evolution is *mahat*, the intellect. It is great in space as well as in time. Hence the name "*mahat*" that means "great one." The individual counterpart of this cosmic state *mahat* is called *buddhi*, the intellect.

From *mahat* evolves *ahamkara,* the sense of I-ness or ego. It is more developed than *mahat*. As differentiation and further differentiation go on in a descending order, the effects, one after another, become more and more developed and subtle. The sense of "I" separates one's individual self from all others and creates an individual entity.

The five senses of perception (hearing, seeing, tasting, smelling and touching) and the five senses of action (verbalization, apprehension, locomotion, excretion and procreation) and the mind arise from *sattvic ahamkara.*

From the *tamas*ic *ahamkara* the five subtle elements called *tanmatras* are derived. They consist of sound, touch, color, taste and smell. *Rajas*ic *ahamkara* does not give rise to any new evolute. It only excites the other two *ahamkaras* to function.

The ears, eyes, tongue, nostrils and skin are the corresponding physical organs of the senses of perception. The mouth, arms, legs, and the organs of excretion and procreation correspond to the senses of action. These physical organs are not the senses; rather they are given powers by the senses. The sense organs come into contact with the objects and are modified into the shape of the respective objects as they are. This is the only function of the sense organs.

Each of the organs of action has gotten its special function that cannot be performed by any other. Mind is both a sensory and motor organ. The sensory and motor organs can operate only through the mind. The mind, the ego and the intellect are called the internal organs, while the five senses of perception or cognition and the five senses of action are called the external organs. The three internal and the ten external organs are collectively called the thirteen organs (*karanas*) in the Samkhya Philosophy.

In the final stage, from the five subtle elements evolve the five gross elements of ether (*akasa*), air, fire, water and earth with one, two, three, four and five attributes respectively (*Samkhya Karika*, 22 and 28). Thus the gross element of ether (or space) with its attribute of sound evolves from the subtle element of sound. Similarly, the gross element air with its attributes of sound and touch evolves from the subtle element of touch, and so on. Finally, the gross element of earth with all the attributes of sound, touch, color, taste and smell evolves from the subtle element of smell.

Thus the total items are twenty-five, namely, *Prakrti* (cause only), *Purusha* (neither cause nor effect), *mahat, ahamkara*, the five senses of perception, the five senses of action, the mind, the five subtle elements, and the five gross elements. These are the twenty-five basic principles (*Tattvas*) of the Samkhya Philosophy.

The purpose of Evolution

The purpose of evolution is twofold. The **first** is the purpose of *Prakrti*, namely, becoming the object of enjoyment of *Purusha*. But *Purusha* is a passive observer so it cannot directly enjoy *Prakrti*. It indirectly enjoys *Prakrti* when it experiences the pleasurable and painful images presented to it by the intellect. This is possible only when *Prakrti* reveals itself to the *Purusha* in and through its various manifestations. This revelation of itself to the *Purusha* is the purpose of *Prakrti*.

The **second** is the *Purusha*'s purpose of freeing itself from the clasp of *Prakrti*. Though it has no attributes, it ascribes pleasure, pain and other properties of *Prakrti* to its own Self, and thus appear to become affected by them. Due to its constant association with the *Prakrti*, it fails to discriminate its own Self from *Prakrti*. But when discriminative knowledge dawns, it realizes its true nature and withdraws itself forever from the association of *Prakrti*. This is called Liberation.

The Sources of Knowledge

The Samkhya Philosophy accepts only three independent sources of valid knowledge, namely, perception, inference and testimony. The other sources of knowledge like comparison, postulation and non-cognition are included under these three sources, and not recognized as separate sources of knowledge.

Perception is the direct cognition of an object through its contact with a relevant sense. For example, we perceive an object like a table when it comes within the range of our vision.

Inference means knowledge derived through the observed relationship between two things. It is the knowledge of a non-observed thing through a related perceived thing. For example we infer fire on a hill from the perception of smoke on it.

Testimony consists of authoritative statements made by trustworthy persons, and gives the knowledge of objects that cannot be known by perception and inference. The nutritionist's statements regarding vitamins, and the scriptural statements about God and immortality are examples of testimony.

The Ethics of the Samkhya

Life on earth is a mixture of pleasure and pain, and even pleasure ultimately ends up in pain or suffering. The **suffering** is of three kinds. The **first** is the suffering arising out of intra-organic causes like bodily disorders and mental afflictions. The **second** is the suffering arising out of extra-organic causes like other human beings, animals, plants, insects, etc. Instances of this kind are injury caused by another person, murder, snakebite, prick of thorn and so on. The **third** kind of suffering arises out of extra-organic supernatural causes like thunderstorm, cyclone, earthquake, etc.

All the blessings of modern science and technology give us only temporary relief from pain or suffering. They do not assure a total and final relief from all the ills to which our mind and body are subject. What is the **ultimate cause of suffering**? *Purusha* (Self) is ever pure and free. But it experiences misery due to its association with the *Prakrti*. Due to ignorance it identifies itself with the body and the mind, the effects of the *Prakrti*. It is the mind and body that are subject to pain and suffering. It is the lack of discrimination between the eternal Self and the mortal body and worldly objects that is the cause of our bondage and suffering. Once discriminative knowledge dawns, we realize our true nature and attain Liberation. But discriminative knowledge is not theoretical. It is the outcome of the practice of virtue and Yoga. The Yoga system constitutes the practical side of the Samkhya Philosophy; it enumerates and elaborates the practical methods that lead to the discriminative knowledge and thus to Liberation.

Bondage and Liberation

Bondage and Liberation pertain to the intellect. Bondage arises when non-discrimination exists and the intellect goes after worldly experiences, and Liberation arises when constant discriminative knowledge prevails. The attainment of Liberation means just the clear recognition of the Self as a reality beyond time and space, and above the mind and the body. It is absolute freedom from all suffering. It is the ultimate goal of our life.(*See* 3.11 Liberation, below).

The Concept of God

There is a great controversy among the Samkhya philosophers regarding the concept of God. The original Samkhya was monistic and theistic. But the classical Samkhya, perhaps under the influence of Jainism and early Buddhism, became atheistic. The earliest available text, the *Samkhya Karika* does not discuss the existence of God. This led scholars to believe that early Samkhya did not accept the existence of God. They argued against the existence of God. There is no doubt that the world must have a cause. But God or Brahman cannot be the cause of the world. God is said to be the eternal and immutable Self; and what is unchanging cannot be the active cause of anything. In the presence of *Purusha*, the ever-changing *Prakrti* is enough for creating the universe. However, later developments in the Samkhya clearly indicate the acceptance of the existence of God. The later philosophers point out that in metaphysical discussions, it is very difficult to explain the nature of the universe and of *Purusha* without accepting a Supreme Being. The existence of God is supported by reason as well as by the scriptures.

THE YOGA PHILOSOPHY

Introduction

Yoga is as old as humanity. It was practiced by Yogis and sages since ages. Its value is lauded in the Indian scriptures of the *Vedas* and the *Upanishads*, and in the Epics and *Puranas*. The importance of Yoga as a means of realizing the ultimate Reality has been recognized by all the important Indian systems of philosophy. Yoga Philosophy is an invaluable gift of the great sage Patanjali who codified the system in his *Yoga Sutras.* Vyasa's *Yoga Bhasya* is the oldest commentary on the *Yoga Sutras*. Bhojaraja's *Virtti* and *Yoga Maniprabha* are very simple and popular works on the Yoga Philosophy. Vijnanabhiksu's *Yogavartika* and *Yogasara Sangraha* are other useful manuals of the Yoga system.

Yoga is not a theory, but a practical method for attaining Self-realization. Yoga means the cessation of mental modifications, and the union of an individual soul with the Cosmic Spirit or Consciousness. It is closely allied to the Samkhya system. The Yoga Philosophy mostly accepts the epistemology and metaphysics of the Samkhya Philosophy, but unlike the Samkhya, it believes in God.

Eightfold holistic System

The Yoga system prescribes an eightfold holistic system of practice for controlling the modifications of the mind, and for attaining Self-realization. The eight steps are: *Yama* (restraint), *Niyama* (observance or moral code), *Asana* (physical posture), *Pranayama* (regulation of *prana*), *Pratyahara* (withdrawal of senses), *Dharana* (Concentration), *Dhyana* (Meditation) and *Samadhi*

(Absorption in Super-consciousness). The individual seeker ultimately realizes his true Self or pure Consciousness. Practicality is the main characteristic of Yoga Philosophy. (For details, see Section III, below).

Samkhya and Yoga: Similarities

The Samkhya Philosophy and the Yoga Philosophy are very much inter-related, and do not vary in their essentials. The Yoga is the application of the theory of the Samkhya in practical life. It mostly accepts the epistemology and metaphysics with 25 principles (*See* **The Samkhya Philosophy**, *above*). The Samkhya lays greater stress on the discriminative knowledge as the essential condition of Liberation, whereas the Yoga lays down a systematic practical method of attaining the discriminative knowledge. By practicing Meditation and its prerequisites, the mind is progressively stilled and purified and thus becomes fit for experiencing the transcendental Spirit or Super-consciousness. Both the Samkhya and the Yoga may be considered as the concave and convex sides of the same sphere.

According to the Samkhya, the sources of knowledge are perception, inference and valid testimony. The Yoga accepts these means of knowing. The *sutra* I.7 of the *Yoga Sutras* says: "Direct perception, inference and revealed authority are the valid proofs."

Differences

The Concept of God: The main difference between the Samkhya and the Yoga lies in their treatment of the concept of God. The earliest available text *Samkhya Karika* of the Samkhya system does not discuss the existence of God. The absence of such discussion led scholars to believe that early Samkhya did not accept the existence of God. However the later developments in the Samkhya clearly indicate the acceptance of the existence of God.

The Yoga, on the other hand, gives a prominent place to *Isvara* or God. *Isvara* is described as a special *Purusha* not affected by afflictions, actions, their fruits and the domains of their accumulated impressions. (*Yoga Sutras*, I.24) For details *see* 2.9 God, below).

The Yoga system also regards devotion to God or surrender to God (*Isvara-pranidhana*) as a means for quickly attaining *Samadhi* (I.23). Surrender to God is also considered as one of the three preparatory requirements for the practice of Yoga, the others being purification (*Tapas*) and self-study (II.1). It is also one of the *Niyama*s to be observed for the practice of Yoga (II.32).

The Twenty-three Evolutes: According to the Samkhya, in the process of the evolution of the universe, the *mahat* (intellect) is the first product of *Prakrti*. From *mahat* arises *ahamkara* (ego); from the *sattvic* aspect of *ahamkara* evolves the mind, the five senses of perception and the five senses of action; and from the *tamas*ic aspect of *ahamkara* evolves the five subtle elements, and from the subtle elements the five gross elements.

The Yoga Philosophy gives alternate explanatory titles to the above twenty-three evolutes as follows (*Yoga Sutras*, II19):

Prakrti	: *Alinga*, without a distinguishing mark
Mahat	: *Linga-matra*, merely a mark
Ahamkara and	: Six *avisesas* or non-specific intermediate links
5 subtle elements	: between *mahat* and the final sixteen evolutes Mind, 5 cognitive senses, } 5 5 active senses, 5 gross } *Visesas* or specific final sixteen products elements.

It is generally thought that the idea of evolution is a new contribution made by science to modern civilization, and Darwin is considered to be the father of this idea. This is not really correct. The idea of evolution has come down to us in one form or another from the earliest times. Patanjali has put the fundamental idea underlying the theory of evolution in one *sutra*: the contact of *Purusha* with *Prakrti* is the means of realizing by the *Purusha* of his true nature, and of **unfolding the powers** latent in both. (*Yoga Sutras*, II.23). He has brought into this generalization the most important aspect of evolution, which is missing in the modern scientific theory.

According to the modern theory there is an evolutionary transformation in the physical forms of species in their effort to adapt themselves to the changing environment. The mere evolution of form would be a meaningless process without an evolution in life. Yoga emphasizes the total evolution that leads to the gradual development of consciousness and to the increase in the efficiency of the vehicle as well. On studying the kingdoms of vegetables, animals and human beings, we find that there is a remarkable increase in the degree of development as we pass from one kingdom to another. Along with that we find that the vehicle also becomes more and more complex and efficient for the expression of the unfolding consciousness. When we compare the brain of a snail with that of a monkey and again with that of a highly civilized human being, we find that there is a tremendous development in the capacity of the vehicle to manifest the latent powers in the consciousness. The mental and spiritual powers exhibited by a highly intellectual person of today are nothing when compared to the psychic powers of an advanced Yogi. It is in the invisible realms of the mind and the spirit that the evolution produces its most magnificent results and the astounding powers of the *Purusha* find their main expression.

The powers of *Prakrti* refers to the capacity of the vehicle to respond to the demands of Consciousness. The powers of the *Purusha,* on the other hand refer to the powers of Consciousness to function through the vehicles. As *Purusha* is pure Consciousness and Consciousness is eternal, there cannot be any evolution of his powers in the normal sense of the word "evolution." But Consciousness has not only to manipulate and control his vehicles with increasing efficiency, but has also to transcend them. This is what Self-realization aims to accomplish.

Mind: In the Samkhya, the mind is categorized into three parts, lower mind, ego and intellect. In the Yoga, the mind is studied holistically and the term *chitta* is used to denote the mind-stuff. It is all-penetrating and it includes the unconscious or lower mind, active mind (*manas*), intellect and ego, that is, the entire inner organ (*antakarana*). All psychical states are the qualities (*dharmas*) of this organ. They arise out of impressions (*samskara*) of the past thoughts and experiences that cling to the psychical organ. *Chitta* undergoes modifications when it is affected by the worldly objects through the senses.

The involvement of the soul in the world of objects and desires is deep and complex. The soul, as pure awareness, has no power of its own to gain insight regarding its involvement or to struggle against its unhappy entanglement in the miseries of the world. The soul is dependent upon the *chitta* (mind), which is part of the body (*Prakriti*), for the effort required to free both mind and soul from suffering (*See* 2.6 Afflictions: Causes and Remedy, *below*). Mind is caught paradoxically in a double predicament. First, in seeking to rid both itself and soul from desire, it must desire to be free from desires. This very desire for freedom from desire becomes an all-consuming desire, and thus the most difficult of all the desires to be overcome. Second, in seeking to free the soul from its association with body, the mind must consciously seek to surrender its own awareness, for when the soul leaves the body, the mind ceases to be conscious. Self-awareness ceases and the mind itself devolves into an unconscious latent lifeless state. Yet, so long as the soul mistakenly identifies itself with body, and so long as the mind mistakenly believes its borrowed awareness as its own, both are deluded. Enlightenment, freedom from delusion, is better for both. The bound soul may undergo reincarnation innumerable times because it remains attached to a subtle body where karmic effects require other lives in which to work out its fate. But through the successful practice of Yoga once the soul experiences Self-Knowledge and thus achieves final Liberation, it is freed from further re-births.

When the soul (*Purusha*) regains its isolation or freedom from all disturbing association with the body, its own integrity is restored. The soul becomes the Self or the ultimate Reality. The freed soul has no memories and no desires and nothing to cause it to become again involved in the miseries of bodily life.

Subtle Body: According to the Samkhya, there are two bodies in a living being, namely, a subtle body and a gross body. The subtle body is an eighteen-fold body and it is made up of the intellect, the ego, the mind, the five senses of perception, the five senses of action and the five subtle elements. It leaves the gross body at the time of death, transmigrates and undergoes a series of rebirths impelled by the karmic effects (the effects of past actions).

In the Yoga Psychology there is no subtle transmigrating body. As *chitta* is all-pervasive, a subtle body is unnecessary.

Suffering: The Samkhya holds that there are three kinds of suffering. These are: (a) *Adhyatmika*: mental and physical suffering arising out of organic causes such as anxiety, depression and other mental and bodily disorders, (b) *Adhibhautika*: suffering and pain arising from extra-organic causes like other human beings, beasts, thorns, etc., (c) *Adhidaivika*: suffering caused by planetary influences and other natural forces such as heat, cold, flood and earthquakes.

The Yoga Philosophy, on the other hand, classifies the causes of suffering or affliction into – (1).Ignorance (*avidya*): considering non-eternal as eternal, not-self as the Self, and the impure as pure, (2) Ego or I-ness(*asmita*): the false notion of the Self as identical with body-mind complex, (3) Attraction (*raga*): liking for things which give pleasure, (4) Aversion (*dvesa*): aversion to things which cause pain, (5) (*abhinivesha*): clinging to life (*Yoga Sutras*, II.3-9).

The Means of Liberation: The Samkhya suggests that the *Purusha* can attain Liberation by means of discriminative knowledge. When the *Purusha* realizes that it is different from the *Prakrti*, and that it is pure Consciousness, then it attains Liberation.

On the other hand, according to the Yoga Philosophy, Self-realization can be attained through the perfected discipline of the will and the control of the mental states through an intensive and regular practice of moral codes, sense control and Meditation. Through the systematic practices of Yoga, the influence of the mind on the soul ceases and its activity can no more ensue. Thus the bond between the mind and the soul is untied. The soul, then, exists alone by itself in its true form, unaffected by any worldly thing and thus is Liberation achieved.

Conclusion

The categories of the Samkhya Philosophy are employed more in the Yoga Philosophy. Hence it is often referred to as the Samkhya-Yoga. The Samkhya Philosophy has its own independent existence, and does not refer to the practical system of Yoga. But the Yoga Philosophy does depend on the main aspects of the Samkhya system, but it gives them a new interpretation. It does not thereby lose its independent existence as a separate system of philosophy, particularly in the sphere of psychology and the doctrine of Liberation.

Questions

1. Describe the nature of *Purusha* and *Prakrti.*
2. How does the Samkhya Philosophy envisage the evolution of the universe?
3. What is the connection between *Prakrti* and *Gunas*?
4. What is meant by bondage and liberation? How do they arise?
5. Give a brief account of the Yoga Philosophy.
6. Examine the relation between the Samkhya Philosophy and the Yoga Philosophy.
7. Compare and contrast Samkhya Philosophy and Yoga Philosophy.
8. Analyze the involvement of the soul in the worldly life and the process of its Liberation.

SECTION II

Patanjali's Yoga Psychology

2.1: Yoga–An Internal Research

An Internal Research

Yoga is practiced for the purpose of realizing one's true nature. It is primarily an internal research for finding answers to such fundamental questions as "What am I?" "What is my real nature?" "Am I a mere mortal being or immortal being?" "Is there God?" "Can I see Him?" "What is my real connection with God?" One cannot find answers to these questions by examining the external world. One has to study his own mind going deep into it. The **power of observation** or attention is the **only method** available for this study. One cannot know his real nature and verify the revelations of ancient sages without developing the power of observing what is going on within. The task is enormously difficult. Yet we have to do it in order to have a real spiritual science. What right do we have to say that we have a soul or God within, if we do not feel it or experience it. If there is a soul we must perceive it; if there is a God, we must see It; otherwise it is better not to believe. As Swami Vivekananda says, "It is better to be an outspoken atheist than a hypocrite" (*Raja Yoga*, p.10).

What is the **stand of people** on this idea? Some people, particularly the so-called "learned" consider that religion and metaphysics are superstitious. Some have faith in God half-heartedly. They think of God only when they face some serious problems. Others have mere blind faith. We cannot blame them for such things, because they are taught to simply believe in the words of ancestors. But a spiritual seeker or **Yogi** seeks the truth, wants to **experience the Truth** for himself. The noble longing for truth of this internal researcher is as real as the longing for truth of a modern scientist.

The science of Raja Yoga places before us a practical and scientifically worked out method of reaching this truth. It gives us a means of observing the internal states. The instrument is the mind itself. The power of attention, when properly focused toward the internal world, will analyze the mind and illumine facts for us. The powers of the mind are like the rays of sun's light, when focused or concentrated, illumine. This is our only means of knowledge. Everyone is using it, both in the external and in the internal world. However, for the internal research, the same minute observation has to be directed to the internal world, and this requires a great deal of practice. To turn the mind inward and focus all its powers upon the mind itself in order to analyze itself is very difficult indeed, because from childhood onward we have been taught to pay attention only to external things and so most of us have nearly lost the faculty of observing the internal mechanism. Yet that is the only scientific approach to the internal search for the ultimate Reality.

All the knowledge in the world has been gained only by the concentration of the powers of the mind. There is **no limit to the power of the mind**. The more concentrated it is, the more power is brought on one point. This is the secret. It is easy to concentrate the mind on external things, as the mind naturally goes outward. But it is not so in the internal research where the subject and the object are one and the same; **the mind studies the mind**. "What has to be done by the Yogi is this: "The powers of the mind should be concentrated and turned back upon it. Then as the darkest places reveal their secrets before the penetrating rays of the sun, so will the concentrated mind penetrate into its own innermost secrets. Thus we shall come to the basis of belief, to the real religion. We shall perceive for ourselves whether or not we have souls, whether or not life lasts for five minutes or for eternity, whether or not there is a God. All this will be revealed to us" (Swami Vivekananda, *Raja Yoga*, p. 14).

Thus the mind is the only instrument for the practice of Yoga. Therefore Yoga Psychology deals with the mind, its modifications and three forces (*gunas*) that pervade the mind.

Since ages saints and sages have experienced the Ultimate

Reality through their inner search, intense practice of Yoga. What has been possible for them is possible for everyone.

Questions

1. Is Yoga an internal research? Substantiate.
2. Why is internal research very difficult?

2.2: Analysis of the Mind–I

The Nature of Mind (*Chitta)*

The mind is an internal instrument (*antakarana*). It serves its master Purusha or Atman or Self within as its instrument. By means of the senses (*indriyas*) or the brain centers the mind operates the visible organs like the eyes, the ears, the nose, etc.

The mind consists of **several aspects**: the desiring faculty (*manas*), the intellect (*buddhi*) or discriminative faculty, the ego (*ahamkara*) or the I-sense, and the bed of memory (unconscious or subconscious).

The ***manas*** is a dynamic aspect of the mind. It gets attracted to outside things through the senses. The ***buddhi*** wills and discriminates. It is this faculty that decides what is right, what is wrong, what is good and what is bad. The **ego** is the basis of human life as a separate individual. It is this "I" that arrogates as "I know it," "I like it," "I own it," etc.

The **bed of memory** is a storage. In it all activities of the mind are recorded as fine impressions (*samskaras*). They come to surface when appropriate environment arises.

Although the mind consists of the above four aspects, it is actually **one unit** with several functions just as one individual with several roles. He is a manager in his company, husband to her wife, father to his children, and a son to his parents, and a friend to his friends. When the mind thinks or doubts, it is *manas*; when it wills and discriminates, it is *buddhi*; when it arrogates, it is *ahamkara*; when it stores, it is a bed of memory.

The Instrument of knowing

The mind with the senses is the instrument of knowing. When senses come into contact with external objects, the mind and the sense are colored by the objects, i.e., they identify themselves with the objects. The conscious mind (*manas*) receives the sense-impressions and forwards them to the buddhi for evaluation. After evaluating the sense-impressions, the *buddhi* assorts, accepts or rejects them. The *ahamkara* arrogates all to itself. When ego's interests are involved in a process of evaluation, the intellect makes a quicker evaluation in favor of the ego. The subconscious mind serves as a store- house of the impressions that are processed by the first three aspects of the mind. It is responsible for recollection, memorization and mental retention.

Thought-waves (*vrittis*)

The activities of the mind arise as ripples in the lake of the mind. Therefore, they are called *vrittis* or waves in Yoga Psychology. All our experiences are sustained by different types of **thought-waves**. All of them are the result of the fourfold aspects of the mind. The unchanging *Purusha* (Self) witnesses the changes in the thought-waves of the mind. But those changes do not cause any change in the *Purusha.*

If the subconscious mind is filled with negative impressions arising out of the functions of the subconscious mind, the intellect and the ego, it taints the ego or renders it unhealthy. A diseased ego creates abnormal attachments, craves for infatuated relationships, develops an intense desire for more and more material possessions, and it indulges in astentation, hypocrisy and selfishness. It vitiates the functions of the intellect, and the functions of the mind and may lead to criminal inclinations. Hence it is necessary to purify the ego through the practice of Yoga disciplines.

The States of the Mind

The **mind**, like other things of the Nature, is **subject to the interplay of three *gunas*** (modes or forces of Nature): *sattva, rajas* and *tamas* (See 2.5 Three *Gunas*, below). When *sattva* predominates, the mind is calm and peaceful; when *rajas* predominates, it is restless and agitated by many desires; and when *tamas* predominates, the mind is dull and passive.

The habitual state or condition of the mind is of five kinds: restless (*kshipta*), stupefied *(mudha)*, distracted (*vikshipta*), one-pointed (*ekagrata*) and controlled or arrested state (*nirudha*).

Restless State: When *rajas* is active, the mind becomes restless. Then it is not fit for concentration on any subtle object and consequently cannot comprehend any subtle principle.

Stupefied State: When *tamas* rises, the mind is stupefied and infatuated. It tends toward sinful activities. It manifests drowsiness, laziness and inertia. In this state the mind is unfit for contemplation.

Distracted State: In this state the mind may be calm sometimes and disturbed at other times. Most of the spiritual practitioners have this type of mind. Therefore, they are in need of regular sustained practice over a long period of time.

One-pointed State: This is the purified state of the mind of the Yogis. The practice of meditation is meant to bring about this state. In this state the mind is focused on the chosen object of concentration. When this state is perfected one attains Samprajnata Samadhi (*See* **3.10 Samadhi**, below)

Arrested State: In this state the thought process is stopped at will by long practice. By this process when the mind gradually ceases to function, Liberation is attained.

The first three states belong to the worldly-minded people, and the last two states belong to the Yogis.

Questions

1. What is the nature of the mind? What are its different aspects and their functions? Does this mean that there are different units of the mind?
2. How does the mind serve as the instrument of knowing?
3. What is a diseased ego? What are its effects?
4. Describe the states of the mind, and their relevance or non-relevance to the practice of Yoga?

2.3: The Modifications of the Mind and their Control

The Waves of Thought (*Vrittis*)

The mind may be compared to a lake. If the lake's surface is agitated by waves and the water is muddy, we cannot see its bottom. But if the water is clear and there are no waves, we shall see the bottom. It is the same with the mind. Below the bottom of the mind is our true Self (Soul). When the **waves of thoughts agitate the mind, we cannot see our Self**, which is our true nature. In the forms of desires, expectations, disappointments, reactions to external events, imaginations, memories, experiences of pleasure and pain, etc., thought-waves go on continuously rising in the mind.

When an external object or event bombards a sense of perception (eyes, ears, nose, tongue or skin), which acts through its corresponding nerve center, an image is formed in the *manas*. The manas takes the impression further in and presents it to the determinative faculty (*buddhi*), which reacts. Then the ego reacts. Then the soul reflecting the mind perceives the object. The soul is the only sentient being; the mind is just an instrument through which the soul perceives the external world. This is how thought-waves rise in the mind and disturb its natural state. Besides, due to the interplay of the *gunas*, the state of the mind goes on changing. When *rajas* or *tamas* predominates, the mind is disturbed. In the *sattva* state only, the waves do not rise, and the mind-lake remains calm and clear. But it is not inactive, because restraining requires greater strength and power. The mind is always trying to get back to its natural state of calmness, but the senses draw it out. The restraint of the mind or checking its outward tendency and focusing it towards the Self is the primary step in Yoga (*Yoga Sutras*, I.2).

Man's Real Nature

You are neither the body nor the mind; you are essentially the eternal Self within the mind. The Self (Purusha) is always seeing your mind and body. So it is the Knower or Seer. It knows the mind's activities or thought-waves or modifications. When you control your mind and restrain its modifications, the Self (Seer) abides in Its true nature, which is eternal peace (*Y.S.*, I.3).

When the mind is completely free from thought-waves or modifications it becomes as calm as a still lake with no waves. The mind, then, reflects the true Self. A perfectly smooth and clean mirror gives the true reflection of your face. So also the clean and calm mind free from modifications, reflects the true nature of the Self within.

If the lake of the mind has a lot of waves (modifications) dirty or colored like the surface of muddy and colored water of a lake, you will be seeing a distorted reflection. To see a clear reflection the water must be clean and calm and without any ripples. Similarly, when the mind ceases to create thought waves or modifications, it becomes as clear as a still lake and you see your Self

At other times, i.e., when the modifications of the mind are not restrained, the Self appears to assume the forms of the modifications of the mind (*Y.S.*, I.4). When you are not aware of your real nature or identity, you may identify yourself with your body. Then you think that you are a man/woman. Or you may identify with your profession and think that you are a doctor or professor, and so on. As a result you undergo pleasure and pain and are subject to the bondage of your karmas (effects of your actions). The same is true with other beings. Forms and names differ, but behind them is one unchanging Consciousness or Self (Spirit).

That is why, we should calm our minds and realize our true identity, Self. Then we shall find the unity among all beings, for the Self in one is the Self in all other beings. Note what the *Isavasya Upanishad* declares,

> "The Self moves, and moves not. It is far and is near; it is within all this, and it is outside all this" (Verse 5).
> "When he, who sees all beings in his own Self and his own Self in all beings, has no fear from anything." (Verse 6).

Lord Krishna endorses this in the *Bhagavad Gita:*

> "The Yogi absorbed in his Yoga sees the Self abiding in all beings and all beings in his Self. He sees the same everywhere" (6.29)

When this unity of all beings and all persons is realized, there can be no conflict in the name of sex, color, caste, race, religion, or place. Therefore the mental modifications are the root cause for all conflicts and problems in the world. **Change the state of the mind** and **make it calm** and serene, then **everything else changes**. When this transformation is accomplished we will find a new world, a world of harmony and peace. This is the purpose of Yoga.

The Kinds of Modifications

The sage Patanjali classifies the modifications of the mind into five kinds. Some of them are **painful** and cause misery and suffering. Others are **painless** (*Y.S.*, I.5). The **painful modifications** are **based on the five causes of afflictions** (*klesas*): ignorance, egoism, attachment, aversion and clinging to life. These five causes cause suffering. The impressions of these modifications produce an afflicted state of the mind and lead to the accumulation of karmas.

The **painful modifications** are selfish thoughts. They ultimately bring pain. For example to love something or somebody is initially pleasurable, but it brings you later a lot of unhappiness, pain, frustration, hatred, etc., because it is based on some expectation in return.

On the other hand the anger of a selfless person may cause you to feel bad in the beginning, but ultimately it helps you to correct yourself and then to lead a better life. We should, therefore, analyze all our motives and try to cultivate selfless thoughts.

When the discriminative enlightenment destroys ignorance and other causes, the mental modifications of the discriminative insight counteract the operation of the *gunas*. Then they do not cause any misery and are, therefore, known as **painless modifications**.

The **five kinds of modifications** are **right knowledge, misconception, verbal delusion, sleep and memory** (*Y.S.*, I.6).

Right Knowledge (*Pramana*)

Direct perception, inference and reliable testimony are the sources or proofs of right knowledge (*Y.S.*, I.7).

Seeing something yourself is **direct perception**, if there has been nothing to delude the senses or if it does not contradict any past right knowledge.

Inference means knowledge derived through the observed relationship between two things. It is the knowledge of a non-observed thing through a related perceived thing. For example we infer fire on a hill from the smoke seen on it.

Testimony consists of authoritative **statements made by trustworthy persons**, and gives the knowledge of objects that cannot be known by perception and inference. The nutritionist's statements about vitamins, and the scriptural statements about God or Self-realization and immortality are examples of testimony. Holy scriptures, whether from East, West, Middle East, Far East or North, contain the words of sages, saints and prophets who have seen the truth and have expounded it. The words or modes of presentation and illustrations may vary depending on the local customs and usage, but the underlying truth of scriptures is same. It is something like a person wearing different outfits in different occasions and different places. The outfits vary but the person wearing them is the same. In the same way the truth of all scriptures is the same, though presentation varies.

Misconception (*Viparyaya*)

This is **false knowledge** of an object and it is not based on its real nature (*Y.S.*, I. 8). Misconception is a wrong knowledge derived by **mistaking of one thing for another**. For example a piece of a mother of pearl is mistaken for a piece of silver.

Verbal Delusion (*Vikalpa*)

This is an **impression formed** on hearing mere words **without** any corresponding **reality** (*Y.S.*, I.9). On hearing a word we do not wait to consider its meaning, we jump to a conclusion immediately. If I say, "I'm a dumb fellow; I can't talk to you;" if you hastily conclude that I am dumb, it is verbal delusion, because if I am really dumb, how can I tell it to you?

Sleep (*Nidra*)

It is a modification based on the **feeling of nothingness** (*Y.S.*, I.10). We think that we have no thought, no experience of right knowledge, misconception, imagination or memory in the mind during sleep. But when we awake, we know that we were sleeping. If there were no thought during sleep, we would not remember that we had slept. In daily life one is asleep to many things due to lack of alertness.

Memory (*Smriti*)

It is **remembrance** of an object or pleasure or pain **previously experienced**. Its impression in the subconscious mind comes back to the conscious mind.(*Y.S.*, I.11). When we perceive an object, an impression is formed in the subconscious mind and at a later time it comes back to the surface. Memory can come from direct perception, misconception, verbal delusion or sleep. Dream is a memory that comes to the surface during sleep.

The above five kind of thought-waves must be controlled to get back the mind's natural state of peace. How can we control these thought forms?

The Means of Controlling the Mental Modifications

The above modifications, says Patanjali, are restrained by **practice** and **non-attachment** (*Y.S.*, I.12). **Practice** is a **positive approach** to thought control, and **non-attachment** or detaching the mind from the cause for modifications is a **negative approach** to thought control.

The river of mind can flow in two directions. It can flow towards

Liberation through the path of discernment. But if it flows towards the objects of the senses through the path of ignorance, it is flowing towards bondage of karmas. In general the river of mind tends to flow towards the objects of the world. This tendency of the mind is weakened by the cultivation of detachment. By the practice of discernment its flow is directed towards enlightenment. Thus the control of the modifications of the mind depends on these two means.

Practice

Continuous effort to keep the modifications **restrained** is practice (*Y.S.*, I. 13). In the restrained state the mind has no modification and it remains peaceful. The effort required for attaining that state is practice. It is an attempt to prevent the mind from going into thought-waves. What sort of practice is required? Should it be casual or for a few minutes a day or now and then? No. Patanjali prescribes **three qualifications**: practice becomes firmly grounded when it is done **continuously** for **a long time** with **great devotion** or **total earnestness** (*Y.S.*, I.14). Thus the conditions are:

1. Practice should be done for **a long time.** We cannot expect to get result immediately like little children who sow a seed today, and dig it up next day to see how much the root went down.
2. Practice should be done **continuously**, not "off and on."
3. Practice should be done with **intense devotion** or **earnestness**, which is acquired through austerity, continence, knowledge and faith.

Note what Patanjali says. He says, "For a long time." He does not say how long. The practice should be **continuous**, i.e., **without break**. And the last qualification is "in all earnestness." That is, **full attention, total devotion** and **faith** and **constant effort** are required to achieve success. Here is a **story** from Hindu scriptures narrated by Swami Satchidananda in his commentary **to illustrate** the importance of these qualifications:

In the heavenly plane where celestials live, there is a great sage called Narada. He travels all over and sometimes comes to earth to see how we are doing. One day he was passing through a forest and saw a Yoga student who had been meditating so long that the ants had built a anthill around his body. The Yogi looked at Narada, and asked, Maharishi (great sage), where are you going?"

"To heaven, to Lord Siva's place."

"Oh, could you please do a favor for me there?"

"Sure, what can I do?"

"Could you find out from the Lord for how many more births I must meditate. Please find out."

"Sure."

Then Narada walked a few miles further and saw another person, singing and dancing with all joy: "Hare Rama Hare Rama Rama Rama Hare!"

When he saw Narada he asked, "Maharishi! Where are you going?"

"To Heaven."

"Oh, that's great. Could you please find out for how long I have to be here like this? When will I get final liberation?"

"Sure, I will."

After many years Narada happened to go by the same route and saw the first man. The Yogi recognized Narada. "Maharishi, I haven't had any answer from you. What did the Lord say?"

"The Lord said you have to take another four births."

"Another ... Four ...births!!! Haven't I waited long enough???" He shouted and lamenting. Narada walked further and saw the second man still singing and dancing.

"Maharishi! what happened? Did you get some news for me?"

"Yes. Do you see that tree there?"

"Sure. I see that tree."

"Can you count the leaves on it?"

"Sure. Do you want me to count them right away?"

"No, no, no. You can take your own time to count."

"But what has that got to do with my question?"

"Well, Lord Siva says you will have to take as many births as the number of leaves of that tree."

"Oh, is that all? So at least it's a limited number then. Now I know where it ends. That's fine. I can quickly finish it off. O Lord, salutations to You."

Just then a beautiful palanquin came down from Heaven, and the driver said, "O Yogi, would you mind getting in? Lord Siva has sent for you."

"Am I going to heaven now?"

"Yes."

"But just now Narada said I have to take many more births."

"Yes, but it seems that you were ready and willing to do that, so why should you wait? Come on."

"And what about the other Yogi?"

"He's not even ready to wait for four more births. Let him wait and work more."

What is the truth behind this story? If you are that patient and love to practice long, your mind is more settled, and what you do will be more perfect and joyful. If you are unsettled and anxious to get the result quickly, you are already disturbed; nothing done with disturbed mind will be perfect. So it is not only **how long you practice**, but with what **patience, sincerity** and **devotion**, with what **earnestness** you do it is important.

Non-attachment (*Vairagya*)

This means **dispassion** of the mind towards the **objects seen** and **heard** of.(*Y.S.*, I.15). The **objects seen** in the world and **heard** from worldly-minded persons are **objects of senses** such as sex, liquor, meat, wealth, family, name, fame, sensual pleasures, etc. The objects heard from the scriptures or religious teachers may be going to heaven after death or experiences of higher worlds or Liberation.

The mind is normally attracted to the objects of senses and tempted by sense-enjoyments. These attractions and temptations create **thirst** for objects and enjoyments, and motivate us to take all kinds of efforts to quench that thirst. The power or **will to resist** such attractions and temptations and to hold the mind in check and thus to prevent it from desiring for sensual objects and desiring for going to heaven is **non-attachment** or dispassion. We should not allow the desires to rule us. This sort of mental strength is called **renunciation**. Renunciation is the only way to freedom and eternal peace.

The Vedantic scriptures say, "Even the desire for Liberation is a bondage." Every desire binds us and brings restlessness. Every desire brings its own color to the mind and influences its thought process. Until that desire is fulfilled the mind is restless. One desire leads to another. If the **mind is tossed** by **one desire after another**, how can it be peaceful? Sri Ramakrishna illustrates this by a story.

All for a single piece of loincloth

A sadhu under the instruction of his *Guru* built for himself a small thatched shed in a quiet place away from villages. He began his spiritual practices in that hut. Every morning after bath he would hang his wet loincloth on a tree by the side of the hut for drying. He would visit the neighboring village to beg for his daily food. One day on his return from the village he found that the rats had cut holes in his loincloth. So the next day he had to go to the village for a fresh one. After a few days later rats again damaged his loincloth spread on the roof of the hut to dry. The sadhu felt annoyed and was wondering, "where shall I go again and again to beg for a loin cloth?"

All the same he visited the village and represented to the villagers the mischief done by the rats. They said, "who will supply you with cloth every day? Just do one thing: keep a cat. It will keep away the rats." The sadhu got a kitten in the village and brought it to his hut. From that day onward the rats ceased to trouble him and he was happy. He now began to tend the useful little creature with care. He was feeding it on the milk begged from the village.

After some days, villagers said to him: "Sadhuji, you require milk everyday. Who will give you milk all the year round? Please keep a cow. You can drink its milk and also give it to your cat." After a few days the sadhu procured a milch cow and there was no need to beg for the milk any more. However, the sadhu found it necessary to beg for straw for his cow. He had to visit the neighboring villages for the purpose; but the villagers said, "there is lot of uncultivated land close to your hut; just cultivate it and you shall not have to beg for straw for your cow."

Guided by their advice, the sadhu took to tilling the land. Gradually he had to engage some laborers and later on found it necessary to build barns to store the grains in them. Thus he became in course of time a sort of landlord. And at last he had to take a wife to look after his big household. He now became a busy householder.

After some time his *Guru* came to see him. Finding himself surrounded by goods and cattle, the *Guru* felt puzzled and inquired of a servant, " a sadhu used to live here in a hut; can you tell me where he has moved?" The servant did not know what he has to say in reply. So the *Guru* ventured into the house where he met his disciple. The *Guru* said to him, "My son, what is all this?" The disciple, in great shame, fell at the feet of his *Guru* and said, "My Lord, all for a single piece of loin cloth!"

Attachment to one thing will lead to another and go on **multiplying** and **plunge you in deep worldliness.** No escape from it. So keep away from attachment to anything.

Verily, attachment has no reference to things. It is all in the mind. The binding link of "I, me and mine" is in the mind. If we do not have this link with the objects of senses, we are non-attached wherever we are and whatever we may be. One may be on the throne and perfectly non-attached, like the king Janaka; another may be in rags and still very much attached.

Non-attachment does not mean carelessness or non-interest in the work you do. You should have greater interest in every work you do. You should do it skillfully with one-pointed attention and devotion and at the same time with perfect dispassion. Then there is no worry, no anxiety. When there is no mental disturbance, you can do a work perfectly and with greater joy. Perfection in action is Yoga.

Only a mind **free from desires** and ripples of thought waves will be **calm** and serene. Therefore **non-attachment** to worldly objects and pleasures is **essential** for the practice of Yoga and peaceful life. It is only with detached mind that we can effectively practice yoga or do a job perfectly or render selfless service effectively. One who is guided by selfishness and personal interest is fogged down by expectations, anxieties, botherations, disappointments, and so will not find peaceful mind to concentrate on the work on hand or Yoga. Sri Ramakrishna was used to tell a **story to illustrate the botheration of selfishness**.

The Root of all Troubles

In a place some fishermen were catching fish. A kite swooped down and snatched a fish. At the sight of the fish, about a hundred crows chased the kite and made a great noise with their crawling. Whichever way the kite flew with the fish, the crows followed it. The kite flew to the west and the crows followed it there. The kite flew to the east and still the crows followed it there. The kite flew to the south and north, but with the same result. As the kite began to fly here and there in confusion, the fish dropped from its mouth. The crows at once let the kite alone and flew after the fish. Thus relieved of its worries, the kite sat on the branch of a tree and thought, "That wretched fish was the root of all my troubles. I have now got rid of it and therefore I am at peace."

As long as you have the "self-fish" – worldly desires, you have to perform actions and consequently suffer from worry, anxiety and restlessness. No sooner do you renounce the worldly desires than your restlessness disappears and you enjoy peace.

A Yogi has to **go further**. The detachment of the mind from its personal desires and enjoyment is **ordinary dispassion** or non-attachment (*Vairagya*). The highest power of dispassion occurs when there is non-thirst even for the *gunas* due to the realization of the Self (*Purusha*) This is **supreme non-attachment (*paravairagya*)** (*Y.S.*, I.16). (*See* 2.5 Three Gunas, below).

While the whole of *Prakrti* (Nature) consists of the three *gunas* or forces: *sattva* (light and serenity), *rajas* (activity and sensuality) and *tamas* (Inertia), Purusha (true Self) alone is beyond all these, beyond nature. While the Prakrti is insentient, the Self is effulgent, pure and perfect. Whatever intelligence we see in Prakrti is but the reflection of the Self upon nature. The Prakrti has covered the Self in living beings, and when it takes away the covering, the Self appears in Its own glory. By practice of Yoga and cultivation of non-attachment, we ultimately realize the true nature of the Self. When the knowledge of Self dawns and we transcend the three *gunas*, we attain the state of supreme non-attachment, and become free from desires. Then we are **established in eternal peace** and bliss. How this state of super-consciousness is attained will be explained in Section III: Patanjali's Yoga System, below.

Questions

1. What are the modifications of the mind? Why should these be restrained?
2. What are the kinds of modifications? Briefly describe each of them.
3. What are the sources of knowledge? Explain each of them.
4. How are the modifications of the mind controlled?
5. What sort of practice is required to control the modifications?
6. Distinguish between ordinary non-attachment and supreme non-attachment.

2.4: Analysis of the Mind–II

Evolution

Patanjali continues to analyze the mind in the fourth Chapter (*pada*) of the *Yoga Sutras* In the process of evolution one species may change into another species. It is generally thought that the idea of evolution is a new contribution of modern science, and Darwin is considered to be the father of this idea. This is not really correct. The idea of evolution has come down to us in one form or another from the earliest times. Patanjali refers to this evolution. He says, **"the change of one species into another is caused by the inflow of nature"** (*Y.S.*, IV.2).

In every species there are body, mind and senses appropriate to its nature. For example the body, mind and senses in a human being are of the human mould, those of an animal are of the animal mould, etc. When there is a transformation of one species into another, the mould appropriate for the change enters the new form and shapes the organs accordingly. The mould of all possible modifications in each of the organ is inherently exists. When self-effort is sustained by a species, Nature begins to endow it with all that is needed for the desired change. A weak nervous system can be transformed into a strong system, a weak mind into a strong mind, and so on. This is implied in the expression "inflow of the nature." How does this inflow take place?

The water for irrigating a field is already in the feeding channel running along it, but water is held back by the ridge (bund) of the channel. What the farmer has to do is to make a breach in the ridge. Then water rushes automatically into the field. In the same way all progress and power are already in man. When good tendencies remove the obstacle of bad tendencies enveloping him, the innate nature of the body and organs brings out the appropriate transformations in them. It is the nature that is driving us towards perfection. All **practices** and all efforts to become pure are not the direct cause for evolution, but **only breakers of obstacles** to evolution (*Y.S.,* IV.3).

This ancient Yogic theory of evolution is better than the modern theory. The cause of evolution, as per the modern theory is survival of the fittest or competition for survival. Competition is only an unnecessary factor caused by ignorance. It is not necessary for the preservation of the human race. The great evolutionist Patanjali declares that the true goal of evolution is the manifestation of the perfection, which is already in every being. Even when all competition ceases, the nature will make us go forward. In the animal there is the potential man suppressed, but as soon as the door is opened man rushes out. So too, in man there is the potential God, covered by the veil of ignorance. When knowledge tears off this veil God manifests.

Creation of Bodies and Minds

The mind is an inexhaustible storehouse of power. A Yogi can learn the secret of its power. He can manipulate it in any way he likes. Mind is a material. Egoism, which is a part of the mind, is also a fine material. When a Yogi finds the secret of these energies of nature, he can create many bodies or minds out of his pure egoism. Do the created bodies have just one mind or do they possess separate minds? They have separate minds (*Y.S.*, IV.4). The effects of one's actions bind him. But one who has attained a state of Liberation does not assume created minds for the purpose of enjoyment or for destroying the effects of his past actions

Though the activities of the created minds are different, one efficient basic mind directs them in order to achieve unity of purpose (*Y.S.*, IV. 5). The artificially created minds (born of meditation) are free from karmic impressions (*Y.S.*, IV.6).

Psychic Powers

When concentration, meditation and Samadhi are practiced together, it is called *Samyama.* By the constant practice of *Samyama*, a Yogi's intellect is freed from the influence of *rajas* and *tamas*, and thus it becomes luminous and bright with the light of intuition. Through this **intuitive light** the Yogi can acquire various psychic **powers** enumerated in the third Chapter of the *Yoga Sutras* (III.16-56).

Some people get psychic powers by other means – out of birth or through medicinal herbs, mantras or austerity (*Y.S.*, IV.1).

By birth: Due to the practice of Yoga in previous births, a Yogi may be born with psychic powers. Many sages were endowed with mystic powers even in their childhood. The divine incarnations such as Rama, Krishna and Jesus, and various sages like Ashtavakra, Vamadeva, Jnaneshwar and others manifested amazing psychic powers.

Medicinal herbs: Ayurveda and Siddha Systems of medicine (ancient Indian Sciences of Medicine) are familiar with the powers and secrets of herbs. Some herbs coupled with mental discipline induce some psychic powers.

Austerity: Many Yogis and sages from ancient times have attained great powers by practicing various austerities.

Thus minds attain powers through herbs, mantras, austerities and Samadhi (meditation). Of these, the **mind** that has **attained** powers through meditation is free from karmic impressions (*Y.S.*, IV. 6). Those who have attained powers through birth, medicinal herbs, mantras or austerities still have desires, but one who has attained to Samadhi through intense meditation is free from all desires. So he is free from karma, and he will not be born again (*See* 2.7 The Doctrine of Karma, below).

The Perception of an Object by Several Minds

Several minds may perceive a single object. Do their perceptions of that object are the same or different? If they are different, what is the reason? Due to **differences in the conditions** of the various minds, their **perceptions** of the **same object may vary** (*Y.S.*, IV.15).

The same object differently. Take for example of a beautiful married woman. She brings joy to her husband, causes other women to be jealous of her beauty, arouses lust in the lustful, and is regarded with indifference by the man of self-control. Which of these observers knows her as she is? None of them. An object in itself cannot be known by the perception of a sense.. Its existence does not depend on a single mind. If it were so, what would happen to it when that mind did not perceive it? (*Y.S.*, IV.16). An **object is not a creation of a mind**. If one mind does not perceive it, other minds, which come into contact with it, will perceive it. Therefore an object has its own existence, and the minds are also distinctive and each individual's mind is peculiar. An object has two aspects: (1) when it is being perceived by a mind; (2) when it exists as an object. The latter is its objectivity. The **perception** depends on and **varies with each individual perceiver.** Sri Ramakrishna tell a **story to illustrate** this nature of the mind.

Several people were walking early in the morning. They saw a man lying by the side of the road. The first person who passed him said, "He must have spent the whole night in the gambling den and couldn't make it home, and he fell asleep along the way. Gamblers are always like that."

The next person to come by said, "Poor man, he must be really sick. Well, probably it's best not to disturb him." And he walked away.

The third person said, "Humph! You dirty fellow, you did not know how much to drink. Someone probably gave you some free drink. You drank a lot, and now can't even get up."

The last person said, "To a saint, nothing matters. Even if he is lying on the payment, he'll just be communing with God. Well, probably, he is above this physical consciousness. Let me not disturb him." And he bowed to the man and walked away.

All four saw the same person, but each saw him differently, because **each one projected his mind**. Each one's projection is based on his thoughts and mental attitude. If there is hell in one's mind he won't see heaven anywhere. If there is heaven in his mind, even a hell will be a heaven for him.

Objects and the Mind

Objects attract and modify the mind of an individual just as a magnet attracts a piece of iron. When the mind comes into contact with an object through the corresponding sense, the object is known to the mind. If it does not come into contact with an object, it is unknown to the mind. The sources of objects are the external actions of sound, light, smell, etc., which reach the mind through the sense-channels and modify it (*Y.S.*, IV.17).

Unlike **objects** that are **sometimes known** and **sometimes not known by the mind**, the **mind as an object of Purusha** (Self) is **always known to It**. While the mind is always in a state of flux, the Purusha is changeless, colorless and pure. All the modifications of the mind are merely reflected upon the Purusha, just as a magic lantern throws images upon a motionless screen without in anyway staining it. Motion can only be perceived when there is something else, which never changes (*Y.S.*, IV.18).

The mind is just an object of perception as any object it perceives in the external world. But the **mind is not self-luminous**; it is not a light-giver like the sun. It is only a light-reflector like the moon. **Atman alone is the light-giver**, and the mind shines and perceives by the reflected light of the self-luminous Atman. It cannot perceive both itself and an external object simultaneously. While it perceives an external object, it cannot reflect on itself and *vice versa.* This proves that it is not self-luminous (*Y.S.*, IV.19, 20).

Perception of One Mind by Another

In order to avoid admitting the existence of the Atman, if a philosopher postulates that one mind is illumined by another, then the latter will be illumined by a third mind, and so on. Then there would be an infinite regress as in a room walled with mirrors. Moreover, there would be an intermixture of memory, causing utter confusion of memory, because of each of these minds would have an individual memory (*Y.S.*, IV.21). Thus the Buddhists have confused the issue by not accepting the existence of Self (*Purusha*), the reflector of the mind and the intellect. Samkhya and Yoga philosophies uphold the existence of *Purusha* as the master and experiencer of the mind. How is this so?

***Purusha* and the Mind**

The mind stands, as it were, midway between *Purusha* and the external objects. Wherefrom it derive its power to perceive the objects? The pure Consciousness of the *Purusha* is constant. It does not change, nor transmissible. But it appears to transmit its light to the mind, because its reflection falls on the mind, and so the mind seems to be endowed with consciousness and appears to be identical with the *Purusha* (Y.S., IV.22). As the **reflection of** the consciousness of ***Purusha*** **falls upon the mind, the mind gets its power to perceive** the objects from that reflection. A mirror cannot reflect the image of a person standing before it if there is no light. But when a light is brought in, it immediately reflects his image. So also, the mind, by getting the *Purusha*'s reflection, appears to be conscious and perceives the external objects.

The mind is colored by the things, which it perceives. At the same time it is itself an object (knowable) of the Purusha, and thus colored by the *Purusha,* the subject and seer. Thus the mind appears to be both subject and object. Though the mind is unconscious, it appears to be conscious through its borrowed power from the *Purusha.* Thus it behaves like a reflecting crystal and appears to know everything (*Y.S.*, IV.23). Seeing this likeness to consciousness, the ignorant regard the mind itself as the conscious entity. In Samadhi what is perceived is reflected in the mind. That which is so reflected in the mind is *Purusha* (Self).

For whom does the Mind exist?

The latent impressions of feelings and thoughts derived from countless previous births lie stored up in the mind. Yet **it does not exist for itself, but only for the experience and Liberation of the *Purusha***, because like a house it is an assemblage of many components (like desiring faculty, intellect, egoism, etc.) and functions with other things like senses, borrowed power from the *Purusha,* etc. A house is just a pile of materials, until an owner comes to live and enjoy it. So also the mind is a mere collection of desires and thoughts. It can only become purposive by the will of the *Purusha.* The states of the mind like happiness, misery or peace are for the benefit of the Knower, the *Purusha*. The Knower is not a complex assemblage, as He is One without any component parts. That is the real Self within us, all the rest are His objects (*Y.S.*, IV.24).

Ignorant persons do not have any inclination to ascertain their true nature. But one who has attained discriminative enlightenment knows undoubtedly the distinction between the mind and the Atman. He ceases to consider the mind as the *Purusha*. After the distinctive nature of the *Purusha* is realized, all queries and doubts about the Self cease (*Y.S.,* IV.25).

When the mind, which was so long occupied with the experience of objects of senses, is inclined towards discrimination, it takes a different turn and moves towards the path of Liberation (*Y.S.,* IV. 26). However whenever there is a break of discriminative knowledge, the latent impressions of the past come to the surface, and the thoughts of "I, me, and mine" arise. This trend would recur now and then until the latencies are completely attenuated through continuous practice of Yoga and discrimination (*Y.S.*, IV.27)..

The obstacles or the causes of human sufferings (*klesas*) are ignorance, egoism, attachment, aversion and desire of clinging to life. These are burnt out by the fire of true knowledge (*Y.S.*, II.10).(*see* 2.6 Afflictions: Causes and Remedy, below) Then they like the roasted seeds become unproductive. The **latent impressions**, which disturb the mind, are destroyed in the same manner, i.e., **burnt out by the fire of knowledge** attained by intensive meditation. Then they do not cause any fluctuation of the mind (*Y.S.*, IV.28). When the mind ceases to act it reverts from the manifest state and resolves back into its primal cause, *Prakrti*

Questions

1. Explain Patanjali's views on evolution.
2. When a Yogi creates many bodies, do they have just one mind or separate minds? Give reasons.
3. What are the other means from which psychic powers arise? Briefly explain each of them.
4. Several minds may perceive a single object. Are their perceptions of that object same or different? Give reasons for your answer.

5. Is an object a creation of a mind? Give reasons for your answer.
6. Are objects always known to a mind? When is a mind known to *Purusha*? Is a mind self-luminous?
7. Is a mind illumined by another mind? Give your justification.
8. Compare and contrast the natures of the mind and the Purusha.
9. Does a mind exist for itself? Discuss.
10. How are the past impressions of the mind removed?

2.5: Three Gunas

Introduction

The theory of *gunas* (forces or modes of Nature) is one of the fundamental doctrines of Hindu Philosophy. The entire manifested universe is made up of three *gunas: sattva* (illumination), *rajas* (activity and passion) and *tamas* (inertia) (*Y.S.*, II.18). These are corresponding to the scientific terms of vibration (wave motion), action (mobility) and position (inertia). All the objects of the universe including human mind are combinations of these three *gunas*.

Nature of the *Gunas*

Sattva, rajas and *tamas* are separate forces. Yet they function in coordination with one another. The nature of *sattva* is purity, peace and serenity; that of *rajas* is activity, sensuality, desire, attachment and enjoyment; and that of *tamas* is slothfulness, inertia, delusion and stupidity. When one of the *gunas* is dominant, the other two remain subdued.

A *sattvic* person cultivates fearlessness. He is endowed with discrimination between real and unreal. He enjoys peace of mind, inner harmony and tranquility. He controls the senses and performs selfless service. He is humble, pious, generous, merciful, forbearing, loving, caring and sharing. He practices ethical discipline.

A *rajasic* person is full of cravings and desires. He is forced to act for their fulfillment. He becomes attached to those who help him to fulfill his desires, and hates those that stand in the way. He runs after sensual pleasures. He is egoistic. He is ever greedy and restless. The more he acquires, the more passionate and greedy he becomes. Nothing gives him satisfaction. He runs after name, fame and comforts. He loses power of discrimination. He is selfish and attached to actions and their fruits.

A *tamasic* person is thoughtless and ignorant. He is negligent and indolent. He forgets everything. He fights with people if they speak ill of him. He has no sense of proportion, and no sense of balance. He remains in a state of sleepiness and lethargy, and in that state of mind he goes through negative withdrawal. He does not experience joy or delight. He becomes passive and suffers from all the illnesses related to passivity. He is controlled by negative emotions. He is depressed and helpless.

In the first step in the spiritual journey an aspirant should not allow the *tamasic* quality to raise its head. He should prevent his mind from going to the grooves of negativity and passivity.

The next step is to learn not to look for objects of pleasure. Sensual enjoyment robs his inner vitality that is the very basis of spiritual development. Keeping *tamas* and *rajas* under control, he should cultivate *sattva guna*. Without establishing *sattva* as the predominant quality, no progress can be achieved in the practice of Yoga. In the next stage by cultivating intense discrimination, the Yogi should overcome even *sattva,* as *sattva* also binds one by making him search for happiness and worldly knowledge.

Therefore **one should transcend the *gunas* to become fit for attaining Self-realization**. The *gunas* are like robbers who cannot come to a public for fear of being arrested. Sri Ramakrishna narrates a story to illustrate this.

Once a man was going through a forest. Then three robbers fell upon him and robbed him of all his possessions. One of the robbers said, "What is the use of keeping this man alive?" So saying, he was about to kill him with his sword, when the second robber interrupted him saying, "Oh, no! What is the use of killing him? Tie his hand and feet and leave him here." The robbers tied his hand and feet and went away. After a while the third robber returned and said to the man, "Ah, I am sorry. Are you hurt? I will release you from your bonds." After setting him free, the thief said, "Come with me. I will take you to the public highway." After a long time they reached the road. Then the robber said, "Follow this road.

Over there is your house." At this point the man said, "Sir, you have been very good to me. Please come with me to my house." "Oh, no!" the robber replied. I cannot go there. The police will know it."

The world itself is the forest. The three robbers prowling there are *sattva, rajas* and *tamas*. It is they that rob a person of the Knowledge of Truth. *Tamas* wants to destroy him. *Rajas* binds him to the world. But *sattva* rescues him from the clutches of *rajas* and *tamas*. Under the protection of *sattva* the person is rescued from anger, passion, sloth, lust and greed. Further *sattva* loosens the bonds of the world. But *sattva* also is a robber. It cannot give him the ultimate Knowledge of Truth, though it shows him the road leading to the Supreme Abode of God. Even *sattva* is away from the knowledge of Brahman. *Sattva* is the last step of the stairs. Next is the roof. As soon as *sattva* is acquired, there is no further delay in attaining God or Self-realization. One step forward God is realized.

Evolution of the Universe

Patanjali's system of Yoga is based on the Samkhya Philosophy. According to the Samkhya, *Prakrti* (Nature) is both the material and the efficient cause of the universe. It is composed of the three *gunas: sattva, rajas* and *tamas*. The whole purpose of *Prakrti* is to give the *Purusha* experience of pleasure and pain, and Liberation (II.18).

When the *gunas* remain in the equilibrium state the *Prakrti* remains unmanifested and the universe exists only in its potential state. When the *gunas* are witnessed by the *Purusha*, the balance is disturbed and the *gunas* begin to vibrate. This primeval vibration releases a tremendous energy within the *Prakrti*, manifesting the universe in twenty-four stages. The first evolute is *mahat* (great one), the cosmic intellect. The individual counterpart of this cosmic *mahat* is *buddhi*, the intellect.

From *mahat* evolves *ahamkara,* "the sense of I" or ego. From the s*attvic ahamkara* evolve the five senses of perception (hearing, seeing, tasting, smelling and touching) and the five senses of action (verbalization, apprehension, locomotion, excretion and procreation) and the mind.

From the *tamasic ahamkara* the five subtle elements called *tanmatras* are derived. They consist of sound, touch, color, taste and smell. *Rajasic ahamkara* does not give rise to any new evolute. It only excites the other two *ahamkaras* to function.

The ear, eyes, tongue, nostrils and skin are the physical organs of senses of perception. The mouth, arms, legs and the organs of excretion and procreation correspond to the senses of action.

In the final stage, from the five subtle elements evolve the five gross elements (*bhutas*) of ether, air, fire, water and earth.

Patanjali in his *Yoga Sutras* describes *Prakrti* as *Alinga* (that which has no cause), and *mahat* as *Linga matra* (indicator only). He also describes *ahamkara* and five subtle elements as six *avisesas* (non-specific states), and mind, five senses of perception, five senses of action and five gross elements (total 16) as *visesas* (specific products) (*Y.S.*, II19). All these products are the modifications of the *gunas*, i.e., the ***gunas* pervade all evolutes**.

The States of the *Gunas*

The states of the *gunas* are specific (*visesa*), non-specific (*avisesa*), defined or indicator (*linga matra*) and undefined or non-indicator (*alinga*) (*Y.S.*, II.19).

When the universe exists in its potential state, the *gunas* are in perfect equilibrium, and their state is described as undefined or non-evolved. When their equilibrium is disturbed, the universe begins to evolve as mentioned in the previous sub-heading. First, the *mahat* (cosmic intellect) is evolved. The *mahat* is the indicator of the unmanifested *Prakrti* and the Self.

In the next stage, the *gunas* enter into combinations, which form the mind, the five senses of perception, the five senses of action and the gross elements of ether, air, fire, water and earth. These sixteen are the specific changes of the *gunas*.

There are six evolutes that are non-specific. They consist of the ego and the five subtle elements of sound, touch, color, taste and smell. The ego is the cause of the senses of perception and the senses of action, and the subtle elements. The subtle elements are the causes of the five gross elements. The *gunas* pervade all the above changes. They neither cease to exist nor come into existence.

Interaction of the *Gunas*

On account of the mutual opposition of the modifications of the *gunas*, **everything is painful to a discriminating person** (*Y.S.*, II.15). In this world, all experiences that come from outside things are ultimately painful. They may give temporary pleasure, but it always ends in pain. No worldly thing can give everlasting happiness. *Sattva, rajas* and *tamas*, reacting on one another, give rise to tranquil, intense or stupefied experience. The products of the *gunas* are in a state of flux. When one *guna* is stronger, the weaker ones cooperate with it. Thus, by their admixture, the *gunas* produce experience of pleasure, pain and stupor. So all experiences have the aspects of all three *gunas*. Therefore nothing can be pure pleasure only to a discriminating person. It is impossible to get enduring pleasure in this world.

Past, Present and Future

In the phenomenal world there is a continuous sequence from changes from latency to causative activity, from this cause ensues knowledge or sentient state, and the sentient state relapses into a state of potentiality. These changes take place in such quick succession, as a rapidly revolving burning coal looks like a wheel of fire. This is the nature of all objects that are manifestations of the *gunas*.

By the concept of time, all objects are marked as past, present and future. Owing to our limited capacity of perception and subtleness of things we may not be able to perceive them fully. What we perceive is the present manifested state of the objects. Their past and future forms do not exist as a visible substance like the present form. The future form exists in subtle form as a thing to be manifested and the past exists in its form already experienced (*Y.S.*, IV.12–13)

All these three aspects of a thing are objects of knowledge. The past and the future exist in subtle forms. If the effects of actions leading to either experience (of pleasure or pain) or Liberation were unreal, then nobody would be engaged in activities. A cause can only bring out an effect, which is already in existence; it can never produce what does not exist. The *Bhagavad Gita* declares:

> "That which is non-existent can never come into being, and that which exists can never cease to be" (2.16).

The forms and expressions of an object may change, but all those changes have existed and will continue to exist within the object. While the **present** form and expression are **visible**, the **past exists in cognition and memory** and the **future** exists **in** its **potential form**. For example, a potter may wish to make a pot from a clod of clay. The pot can be said to exist in the clod of clay in a subtle form (*Y.S.*, IV.12).

All objects in the universe are composed of the three *gunas*. If all objects are products of the three *gunas*, then how can there be a single perception as cow, tree, hill, etc. The *gunas*, though three in number with three different properties – illumination, activity and inertia, they are inseparable and act in **unison** and produce a change. That is why the product of the change is regarded as **one object.** Since all the three *gunas* exist in every combination, an object preserves an essential unity even in the diversity of its forms and expressions (*Y.S.*, IV.14).

Hence we see that the same mind exists throughout the many rebirths of an individual. It is only the play of the *gunas* that makes the mind change its form and expression in different rebirths: predominantly evil or predominantly good or predominantly passionate and egoistic or peaceful, depending on the nature of karmas that fructify in a birth. In the mind of a good person, the past evil impressions still exist in subtle form.

How, then, is Liberation possible? Patanjali answers this question. The subconscious tendencies have their basis in the mind. Therefore **one must cease to identify with the mind in order to attain Liberation**. When he realizes beyond doubt that he is the Atman, and not the body-mind complex, he becomes free from his karma.

The mind of a liberated soul has no more purpose to serve for him. It is resolved back into the *Prakrti* (Nature). (*See* 3.11 Liberation, below).

What does happen to the *gunas* then? By the persistent practice of meditation, a Yogi finally attains the *Dharma megha* Samadhi (virtue-pouring cloud). Then for him the *gunas*, having fulfilled their purpose, cease to have any further sequence of transformations.

The essence of *Purusha* and the *gunas* is never destroyed. Therefore both are eternal. While *Purusha* is immutably eternal, the *gunas* are mutably eternal. Their mutability never comes to an end; but in various evolutes like *buddhi* through which the *gunas* manifest themselves, the sequence of moments comes to an end. The natural mutation of the *gunas*, however, continues and is experienced by other persons who are still in bondage. **For a liberated person the *gunas* have no more purpose to serve, and so cease to function completely** (*Y.S.*, IV.32, 34).

Questions

1. What are *gunas*? What is the nature of the *gunas*?
2. How does the evolution of the universe take place?
3. Explain the states of the *gunas*.
4. What is the consequence of the interaction of the *gunas*?
5. Do the states of the past and the future exist like that of the present? Substantiate.
6. If all objects are the products of the three *gunas*, then how can there be a single perception?
7. How does the same mind exist throughout the many births of a person?
8. When a person attains Liberation, what does happen to his *gunas* and the mind?

2.6: Afflictions: Causes and Remedy

Introduction

The Philosophy of Yoga is not an armchair philosophy that is purely speculative. It is based on hard realities of life. It deals with the real and deeper problems of human life and provides effective means for their solution. The ancient sages and seers who expounded this philosophy were guided by the spirit of scientific inquiry, and were inspired by Divine Grace.

The philosophy of human suffering or afflictions (*klesas*) constitutes the foundation of the Philosophy of Yoga, whose aim is to give freedom from suffering. In the *Yoga Sutras*, Patanjali analyses the fundamental causes of human suffering and the removal of those causes.

The Causes of Human Suffering

There are **five basic causes** of human suffering. They are **ignorance, egoism, attachment, aversion and clinging to life** (*Y.S.*, II. 3). These causes are not separate; one leads to the next. Ignorance is the root cause of the other four causes. These five obstacles to Yoga are forms of wrong cognition. They strengthen the sway of the three *gunas*. The activities of the *gunas* are deep-rooted. They bring about a chain of changes and a flow of cause and effect, and the fruition of our actions.

The States of the above five Causes

The causes of afflictions may exist in the dormant (*prasupta*), feeble or attenuated (*tanu*), interrupted (*vichhinnna*) or active (*udara*)state (*Y.S.*, II.4).

The **dormant state** is a **potential** state, i.e., the **obstacle exists** in a **latent** form. It awakens when proper condition arises. For example, in children the causes of suffering exist in latent form. They are awakened as the children grow.

The **feeble** or **attenuated state** is one that is **thinned by the practice** of Kriya-yoga (for details, *see* 3.1 Kriya-yoga, below).

In the **interrupted state** a cause is **suppressed by its opposite**, but it may come back again. For example, at the time of anger, aversion is in operation and so attachment is suppressed.

In the **active state**, a cause is **fully operative** without any distortion or restriction. This is seen in the case of average people. Their minds are always affected by illusion or ignorance.

In the fifth **burnt-up state**, the seeds of causes are **burnt up in the fire of discriminative knowledge**, which a Yogi acquires in the advanced stage of the practice of Yoga. The burnt-up causes cannot become active even when appropriate occasion arises. Just as a roasted seed cannot sprout, so the causes burnt up in the fire of knowledge cannot affect the Atman. Then the Yogi becomes *Jivanmukta* (Liberated while alive). He is free from the modifications of the mind (*Y.S.*, II.4).

Ignorance (*Avidya*)

What is the nature of ignorance? It is not a mere lack of knowledge. It is unawareness of the Reality. It is **taking one thing for another**. For example a piece of rope is mistaken as a snake during twilight. Ignorance means taking a transient object as permanent, an impure object as pure, the painful as pleasant and the non-self as Self (*Y.S.*, II.5). It is the breeding ground for other causes.

Ordinary people are not aware of their real nature. They are not aware that they are really the eternal, pure and blissful Atman. They wrongly identify themselves with their bodies and minds, and bemoan the decay of their bodies. This illusion is the root cause of all miseries.

Ignorance is the basis for the flow of all causes and for the corresponding latent impressions stored in the unconscious mind along with their karmic effects.

The body of a living being is impure, because of its germinal source, its secretions, etc., but out of ignorance people describe a young girl as beautiful and tender like the crescent moon. This is the misconceived idea of purity in impurity. People think that they enjoy happiness from worldly objects of their liking. But worldly objects cause suffering in consequence in their afflictive experiences and in their latent tendencies. Instead of true happiness, the ignorant people cling to fleeting pleasure. But alas, their satisfaction is short-lived. Ignorance betrays the people. Yet as they sadly turn away, their eyes fall upon some new object of sensual pleasure. So the cycle of hopeless search, momentary pleasure, pain, disappointment goes on.

In the *Bhagavad Gita*, Lord Krishna points out how the brooding on sense-objects leads to self-destruction:

> "By constant brooding on sense objects, attachment to them arises, from attachment arises desire, from desire anger is born" (2.62). "From anger arises delusion, from delusion loss of memory, from loss of memory reasoning is ruined, and from loss of reason, one is lost" (2.63).

When the mind broods over sensual pleasure, a series of chain reactions sets in. It ultimately ends in self-destruction.

Egoism (*Asmita*)

The ego is the **sense of :I, me and mine**." Egoism is the identification of the power of the Seer (Purusha) with that of the instrument of seeing (*Y.S.*, II. 6). The Seer is really the Self (Purusha) within us. The mind, the intellect and the sense organs are the instruments for seeing the external objects. Out of ignorance the Self is identified with the instruments, and we say, "I am hungry," "I am sick" or "I am happy," and so on. We fail to see that Self (Purusha) is different from mind and intellect. The **Self is infinite and unchangeable**. Nothing can produce an effect on It; yet, **through ignorance, we consider the body-mind complex as the Self** and think we feel pleasure and pain.

Ego or the feeling of "I, me, mine" is the **source of desire** for sense objects and pleasures, selfishness, attachments to desirable things, aversion to unpleasant things, negative emotions such as lust, greed, anger, hatred, jealousy and others, **and selfish actions**. All these impurities vitiate the mind. The mind becomes restless and wandering. There is no peace, but there is ego. The ego makes it impossible to experience happiness in our own being, to experience the Truth.

Once there was a man who heard that he could attain supernatural powers by performing a certain practice. So he did that for years. Many years later he attained the power to multiply himself into thirty identical forms. He was thrilled. He was happy. He went from place to place performing this trick. Everyone appreciated his miracle. This boosted up his ego. He kept on doing this feat.

However great one is, one cannot avoid one thing in life, and that is death. The man's final day came. Because of his practices, he knew that the hour was approaching. When he heard the bells of the messenger of death, he multiplied himself into thirty identical forms. The messenger just stood there. He was supposed to take one person, but here were thirty persons who all looked the same. He was confused. He went back to the Lord of Death and reported, "Lord! there were thirty identical persons. What could I do?"

The Lord of Death knows the weakness of the human beings very well. He called another messenger and whispered into his ear, "Praise him! Praise him to death!"

This messenger went to the man, who again multiplied himself into thirty forms. The messenger began to praise him, saying, "wonderful! You are a great man. You are magnificent. You are incredible. There is no body like you. You are the best. You have surpassed God in your power of creation!"

As the adjectives got more and more superlative, the balloon of the ego kept getting bigger and bigger, and all the thirty heads swelled up. The man was totally intoxicated with the praises of his feat.

Finally the messenger said, "But there is one tiny mistake."

The real person jumped forward and shouted, "What's that?"

The messenger of death put the rope around his neck and dragged him away.

This story highlights the intense attachment of an egoistic person to his body and his intoxicated clamor for praises. **Ego vitiates the very thought process and attitude** of a person and his peace of mind.

Attachment (*Raga*) and Aversion (*Dvesa*)

Things or other persons that make a person feel happy, pleasant and that inflate his ego attract him. He develops a sense of attachment towards them. So **attachment** is **that** which **follows** the **experience of pleasure** (*Y.S.*, II.7). The attraction and consequent attachment to things are the outcome of the latent impressions of pleasure enjoyed earlier.

Having forgotten the direct source of bliss (*ananda*) within, one seeks it in the external world and is attracted to persons or things from that one derives momentary pleasure. Such attraction gives rise to addiction and attachment. The fear of losing or actual loss of things to which one is attached causes disappointment, anxiety and sorrow.

One may dislike persons, things and events that make him unhappy or uncomfortable or threaten his security and happiness. So **aversion** is **that** which **follows the experience of pain** (*Y.S.*, II. 8). It also arises out of pain or suffering experienced earlier. A sustained attitude of repulsion casts into the unconscious mind a subtle stream of impressions of aversion.

The likes and dislikes are similar to the two sides of the same coin. They bind us to innumerable persons and things, and affect our life to an unbelievable extent. Our thoughts, feelings and actions are prejudiced by these biases. They **disturb our peace** of mind and keep us tied down to worldly life.

Clinging to life (*Abhinivesa*)

All living beings have craving to live forever. Even one who has lived for over hundred years and nothing more to gain from life has attachment to life. Clinging to life, which flows by its own potency exists even in the minds of the learned (*Y.S.*, II.9).

Every creature has this kind of craving. How does it arise? Although the Self is really immortal, because of the identification of the Self with the body, the **fear of its death** exists in everyone. It is spontaneous. It arises from the subconscious impressions. One who has not felt the dread of death before cannot have the craving for life, for death has not been experienced in this life. Therefore it is inferred that it was experienced in a previous life. Thus the clinging to life proves the existence of previous life. There can be remembrance of only things experienced or felt before. Every experience, feeling or thought is stored in the subconscious mind. Its recollection is memory. The fear of death is also such a memory.

It may be argued that the fear of death is inherent, and so no previous experience is needed for it. If recollection of death is called inherent, then every memory may be said to be natural. But memory is not inherent. It arises from a cause, i.e., past experience. Therefore the fear of death cannot be called inherent.

Ducklings run to the water as soon as they come out of the eggs. Without previous experience how are they able to swim? You may say it is instinct. What is instinct? A piano player can remember when she started learning, how carefully she had to put her fingers on the black and the white keys one after the other. But after long practice, she could talk to someone and at the same time play without seeing the keys. By practice it becomes instinctive and automatic. So with every work; by practice it becomes instinctive. Every creature has died hundreds and thousands of times in the past and knows well the pangs of death. And so when it gets into a body, it is afraid to leave it.

Instinct is involved reason. As reason cannot come without experience, so all instinct is the result of past experience. Does that experience belong to a particular soul or the body? Modern scientists hold that it belongs to the body; but the Yogis hold that it is the experience of the mind transmitted to the body. This is the basis of the theory of reincarnation or rebirth.

Matter is considered to be without a beginning. Mind is matter and so it is without a beginning. The mind undergoes a series of modifications appearing and disappearing. The mental modifications are caused by the interplay of the three *gunas*. The three *gunas*, being without a cause, the flow of modifications resulting from their changes must also be without a beginning.

The mind's recurring experience of various fears produces the clinging to life. That is why at birth a child is instinctively afraid and cries; it has the past experience of pain. Even a learned person, who knows that the body will die but the soul cannot die, clings to life like an ignorant person, because it has become instinctive. All the past experiences of death become subtle impressions in the subconscious mind.

The Removal of Causes

The above five causes exist in **two states**: the **potential form** before they come to the surface of the mind, and the **manifested active state**. It is easier to control the manifested things. For example, when a person is angry, it is easy to see that aversion is in full operation. He can control it by deep breathing or diverting his attention.

The manifested obstacles are first thinned or weakened by the practice of Kriya-yoga. Then they are reduced to the state of the roasted seed by discriminative knowledge attained through meditation (*Y.S.*, II.11). This is similar to washing a dirty cloth; first, the gross dirt is loosened with soap; then the subtle dirt is washed away with clean water.

When the manifested causes have thus been controlled, they still exist in subtle form as subconscious impressions. These can be completely removed only when the mind's activity ceases. When the Yogi transcends the mind in the higher Samadhi, then his mind, having fulfilled the purpose of its existence, gets absorbed into the Prakrti, its primal cause (*Y.S.*, II.10). (*See* 3.10 Samadhi, below)

Why should the Causes be destroyed?

The above **five causes** (ignorance, etc.,) are the **underlying causes of the karmas** that we generate by our thoughts, desires and actions (*Y.S.*, II.12). Every thought, desire or action produces its corresponding result. Every human soul goes through a continuous series of births and deaths reaping the fruits of past karmas.

During the process of experiencing their effects, new karmas are created, and so this process goes on. Good karmas give pleasant experiences, and bad karmas give painful experiences. But even pleasures ultimately end up in pain, because they are associated with anxiety, disappointment and fear.

Is there no end to suffering?

People, in general, say that life is like that, and that one has to take life as it is, and to make best out of it. The common pious people live in the hope that tomorrow will be better. They resign to the fate and bear the pains with stoic indifference.

The stand of Yoga Philosophy is totally different from these ideas. So long as you are bound by illusions and ignorance, you cannot avoid misery and suffering. By practicing Yoga, you have to rise above the illusions and miseries of life, and to gain eternal peace and bliss through Enlightenment.

The Permanent Remedy

The root cause of misery is the entanglement of the soul with matter (*Y.S.*, II.17). The wrong identification of the soul with the body and mind is due to ignorance or lack of awareness of its real nature (*Y.S.*, II. 24). When you reach the final stage of the practice of Yoga, you will transcend the body and mind, and will get freedom from bondage and suffering (*See* Section III of this text, below).

Questions

1. What are the causes of human suffering?
2. What are the states in which the causes may exist? Briefly explain each of these states.
3. What is the nature of ignorance? What is its outcome?
4. What is egoism? What is its consequence?
5. What are attachment and aversion? How do they bind us?
6. Why do living beings cling to life? What is its consequence?
7. How are the causes of suffering removed? Why should they be removed?
8. Is there a permanent remedy to sufferings?

2.7: The Doctrine of Karma

Introduction

The Sanskrit word *karma* is derived from the verbal root *kri*, to act or to do, and the word is almost naturalized into English. Karma means not only "action" but also the "effect" or the "result" of an action. Our desires, thoughts and actions, being governed by the irresistible Law of Causation, produce reactions. The reactions or consequences of karmas are experienced as pleasure or pain.

All thoughts and actions are stored in subtle forms as imprints or impressions (*samskaras*) in the subconscious mind. The receptacle for these latent impressions is called the womb of karmas (*karamasaya*). The latent impressions wait for an opportunity to come to the surface and bring out their reactions either in the present life or in a future life, because **for every action there will be a reaction or effect.** No action goes without its reaction or effect (*Y.S.*, II.12). Good actions bring good results, and evil actions bear evil results.

While the latent impressions of actions that produce results are called *karamasaya* (womb of karma), the latent impressions of feelings arising out of the consequences of actions are called *vasanas* (subtle desires). For example, one is born as a man as a result of his past actions. He goes over his span of life enjoying various pleasures and pain. The impressions stored in the subconscious mind in the course of his existence as a human being go to form the human *vasanas*, which will become active only when he gets embodiment as a human being. The outcome of *vasanas* is memory. Memory is a modification of the mind. For each modification there is an associated feeling. Therefore each memory of experience of pleasure or pain is shaped by a corresponding latent impression of that feeling.

Vicious actions generate vicious impressions in the form of karmic seeds. These impressions give rise to *vasanas. Vasanas* give rise to corresponding actions and experiences. From these actions new impressions are formed. Only those desires become active for which the environment is suitable. Thus the cycle continues until one ends it by attaining discriminative enlightenment (*viveka kyati*) through the practice of Yoga (*Y.S.*, IV.8). (*See* 3.11 Liberation, below)

Continuity in *Vasanas*

Though separated by time, place and class of birth there is continuity in the *vasanas* or subtle desires, because of association between memory and latent impressions (Y.S., IV.9). Experiences become latent impressions, and impressions recalled become memory. Actions are performed in different births, at different times, in different places. Yet by the force of memory, only those actions that are to fructify come together and give rise to corresponding *vasanas* in the mind. Suppose one is born as a man, and then, on account of evil deeds done, he is born as an animal, say, fifty times, and then he is born again as a man. In spite of the intervention of fifty animal births, the human *vasanas* will come up to the surface when he is born again as a man. This happens due to association between memory and latent impression. Then the animal *vasanas* remain in storage.

Impressions—no beginning

The subconscious impressions have no beginning, because the desire to live is eternal. All experience is preceded by a desire for happiness. Each fresh experience is built upon the tendency generated by past experience. Therefore desire is without beginning. In every creature there is a desire for happiness. It arises from a desire to live, which is eternal (*Y.S.*, IV.10). Can desires vanish?

Disappearance of Impressions

From a cause like virtue comes pleasure or happiness, and from vice comes pain or misery. From happiness arises attachment, and from pain arises aversion. From attachment and aversion ensue efforts or actions through the mind or speech or body. By these actions a person either helps or hurts others. From that again arise virtue and vice, pleasure and pain, attachment and aversion. Thus revolves constantly this six-spoked wheel of existence.

Ignorance, which is the basis of all causes of afflictions is the motive power of this perpetually moving wheel. This process thus serves as the cause. The result is the purpose of an action, which determines its moral value as virtue or vice. Nothing that does not exist can come into being, i.e., the **effect exists in a subtle form in the cause**. The **mind** is the **substratum** of the subconscious **impressions**. Without a supporting substratum the impressions cannot exist. A particular **sense-object** that induces the manifestation of a particular impression is its **support**. Thus the subconscious impressions are held together by **cause** (ignorance), **effect, basis** and **support** (*Y.S.*, IV.11). When these disappear, then the impressions also will disappear (See the last paragraph in this Chapter, below.

The karmas are based on the pain-bearing **causes of afflictions** (*klesas*) – ignorance, egoism, attachment, aversion and clinging to life. As long as these causes exist in the mind so long the soul must continue to go through the cycle of births and deaths in order to experience the effects of karmas.

Kinds of Karmas

Karmas are of **four kinds**: **white** (good), **black** (bad) and **black and white** (mixed) and **neither white nor black.** The action of a wicked person is black. The actions of ordinary people are white and black, because they are done by external means and give benefits to others or hurt others. It is difficult to lead domestic life without benefiting or hurting others.

The actions of those engaged in repetition of *mantras*, austerities, spiritual study and meditation, devotion to God and saints are mental and free from external means. They are white, because causing pain to others is not inevitable in such cases. The actions of an enlightened sage or Yogi are neither white nor black, because of their spirit of renunciation and inherent non-violent nature. A sage performs actions without any sense of doership and selfishness and so no new karmas are generated. Established in the Self, he is beyond virtue and vice, and all his actions are neither white nor black (*Y.S.*, IV.7).

While the actions (which are neither white nor black) of the enlightened Yogis or realized sages do not fructify as they are not karmas at all, the white, black, and mixed karmas of others bring out corresponding good, evil and mixed effects.

The Categories of Karma

You may be performing innumerable actions in your present life. Some of them may bear fruit in the present life itself. Others will remain in the womb of karmas and come into effect in your next or future births. So when you take a birth, you not only experience the effects of past actions but also create new karmas. Therefore there are **three** main **categories of karmas**: *Sanchita* (Dormant or accumulated), *Prarabdha* (Active) and *Agami* or *Kriyamana* (Potential) karmas.

***Sanchita* karmas** are 'stored' karmas. They are the sum total of all the accumulated good and bad karmas of the past. They remain dormant in subtle form in the subconscious mind. They will become active when conditions become favorable. When a spiritual seeker gets illumination through intuitive knowledge and attain liberation, the subtle impressions of the accumulated karmas are burnt up by the fire of Self-knowledge.

***Prarabdha* karmas** are a set of past karmas that have become active. It is these karmas that are responsible for the present life. Their effects are unavoidable and must be accepted with a poised mind. Our pleasure and pains, prosperity or poverty, illness or wellness, etc. are the effects of *prarabdha karmas*.

***Agami* karmas** are the current actions that will fructify in future. They are actions or karmas only to those who are egoistic and selfish and have the sense of doership. *Agami* karmas are not actions to a realized sage, as he has no desires, no selfishness and no sense of doership.

In the Yogic texts there is a beautiful analogy. *Sanchita* karmas are like the bundle of arrows stored in the quiver of a bowman on his back; the *prarabdha* karma is like the arrow that he has shot and already in flight. *Kriyamana* karmas are the arrows that are being made or yet to be made. One has control over the *sanchita* and the *kriyamana*, but not over *prarabdha*. The latter has to be experienced, and cannot be avoided.

The Fructification of Karmas

The latent impressions in the womb of karmas begin to fructify when the causes for afflictions exist in the mind, but not when their roots are cut off. This is similar to the germination of the grains of paddy (rice). The un-burnt grains with the husk are capable of germinating, but the grains from which the husks are removed or the burnt-up grains cannot germinate.

The **fruits** of fructifying karmas are of **three kinds: class of birth** (*jati*), **span of life** (*ayu*) and **experience** or enjoyment (*bhoga*) of **pleasure** and **pain** (*Y.S.*, II.13).

What may be the next birth? Depending on the nature of karmas that fructify one may get a human body or an animal body. If one's thoughts are animalistic, one may get an animal body. Of course within each form exists a soul on its evolutionary path toward Self-realization.

One's span of life may be short, medium or long. This is also determined by the karmas. The pleasant or painful experiences one may get in the next birth are also determined by the fructifying karmas.

The actions done under intense lust, anger, compassion, spirit of forgiveness, etc. are **dominant karmas**. They are capable of bearing fruits independently. **Minor karmas** do not become operative independently, but act as secondary to the dominant karmas. When very strong or primary karmas bear fruits, the opposite secondary ones remain subdued, and they can fructify at some future time if aroused by some kindred karmas. For example, a person performs pious deeds in his boyhood. Then as a young man he commits many beastly acts through greed. At the time of his death the fully matured latencies of sins form the appropriate womb of karma. As a result he gets the life of a beast in the next birth. Then the impressions of his pious deeds do not become operative. They remain stored up and will become operative when he is again born as a man.

Your own Creation

Your present happy or unhappy life is the effect of your fructifying past good or evil deeds. Good actions bring happiness and evil actions cause suffering. Therefore, **your happy or unhappy life is your own creation** (*Y.S.*, II.14). Nobody else, neither your parents nor God, is responsible for it.

The causes of misery are ignorance, egoism, attachment, aversion and fear of death. Consequently actions that are opposed to them are considered virtuous, while actions that support the causes are vicious. Contentment, forgiveness, self-restraint, non-covetousness, purity, control of senses, wisdom, truthfulness, compassion and benevolence are pious acts. Anger, greed, jealousy, violence, incontinence, etc are sinful actions.

Just as misery is undesirable to ordinary beings, so to a Yogi of discrimination everything, even pleasure, is undesirable, because it eventually involves pain due to its consequences: the anxiety and fear of losing what is gained; renewed cravings arising from the resulting impressions left in the mind; and the constant conflict among the three *gunas* that control the mind (*Y.S.*, II.15). For example you may be delighted when you gain wealth or high position and receive the appreciation of your family and good friends. But you will start worrying about the possibility of losing it. Similarly a pleasure that you enjoyed in the past will create renewed cravings for enjoying it again.

Everyone, therefore, must endeavor to cherish good thoughts and perform good deeds without any attachment; everyone must also gain an insight into the art of going beyond good and evil, and aim at overcoming the five causes of afflictions in order to attain Self-realization or Liberation (*See* 2.6 Afflictions: Causes and Remedy, above).

The cycle of birth and death and its end

All actions performed in one birth are not redeemed in the same birth. Unredeemed karmas accumulate. A part of them due to fructify causes the next birth, and fresh actions are performed and further accumulation of karmas take place. Thus the cycle of birth and death and accumulation of karmic effects go on. Is there no end to this cycle? There is an end to this, says Patanjali. When a Yogi experiences *Dharma Megha Samadhi*, the five causes of afflictions and the consequent karmic entanglements are totally destroyed (*Y.S.*, IV.30). The Yogi then becomes established in the Self and he is no longer bound by the wheel of birth and death. He is liberated (*See* 3.11 Liberation, below).

Questions

1. Briefly explain the Doctrine of Karma.
2. Differentiate *vasanas* from *karmasaya*.
3. Is there continuity in the *vasanas*?
4. How does the wheel of existence constantly revolve?
5. What are the four kinds of karmas? How do they arise?
6. Briefly explain the three categories of karma.
7. How do karmas fructify? What are the fruits of the fructifying karmas?
8. Who is responsible for your happy or unhappy life?
9. How is every experience painful to a Yogi?
10. Can desires vanish? Give reason for your answer.

2.8: Purusha and Prakrti

The Samkhya Philosophy and Yoga Philosophy admit two ultimate Realities: Purusha (Self) and the unconscious Prakrti. Both are equally real and eternal.

Purusha

Purusha is ever pure and ever free, but it becomes subject to pleasure and pain when it forgets its true nature and identifies itself with Prakrti. Purusha consists of multiple selves. This multiplicity is proved with reference to the differences in the nature, activity, birth and death and sensory and motor endowments of different beings. But all these differences pertain not to the Self as pure Consciousness, but to the bodies associated with it. According to the Vedanta, there is only One Reality, viz., Atman or Self, and everything else, including Prakrti, is Its manifestation.

Purusha is absolute knower. It is not subject to any condition. It itself is Consciousness. It is free from affliction. It is changeless, immutable. The mind is always witnessed by Purusha, and it is changeable. The mind knows an object when it comes into contact with it through a sense. But it does not know an object when it is not within its range.

Moreover the mind serves the purpose of another, while Purusha is the end in itself. The mind is composed of three *gunas* and is unconscious. Purusha illuminates the mind and thus it appears to be identical with it (*Y.S.*, II.20). Purusha has no attributes. Then how does it appear to be happy or unhappy?

Egoism

The mind expresses itself through the changing *gunas*. The true nature of Purusha is unchangeable, awareness, eternal freedom and immutability. If, due to ignorance, the mind, which belongs to not-self, is taken to be Purusha, the power of Purusha is looked upon as identical with the power of the mind. This confusion results in egoism, which gives rise to affliction (*Y.S.*, II.6)

The Cause of Pain

The experience of pleasure and pain is caused by false identification of the Purusha with Prakrti. In fact Purusha (Self) is eternally free. By its false identification it is mistaken for the individual ego and it is subject to all the thought-waves of the mind. Thus it experiences pleasure and pain, which are avoidable (*Y.S.*, II.17).

Purusha unchangeable

If Purusha also changes its nature like the mind, then the modifications of the mind would be known at times and unknown at times like the objects of the senses. But, in fact, the mind is always known to its master Purusha. This makes one to infer the Purusha's unchangeable nature (*Y.S.*, IV.18).

Purusha: the only Seer

The mind is not self-luminous, because it is an object of perception by the Purusha. All the evolutes of Prakrti are knowables, while the Purusha is the only "seer" or "perceiver." The senses perceive objects by the light borrowed from the mind. The mind, in turn, borrows light from the Purusha, which is self-luminous. The mind is knowable, because from a reflection of the action in one's mind, one feels such thought-waves like "I am hungry," "I am afraid," "I like it." This would not have been possible had there been no perception of what is happening in one's own mind (*Y.S.,* IV.19).

Moreover, if there were no assistance from the Purusha, the mind could not have the perception of itself and the external objects at the same time. But in practical life it is seen that a person is aware of the mental modifications and the sense-objects at the same time.

This proves that the Purusha, which is distinct from the mind, accomplishes this simultaneous perception (*Y.S.*, IV.20).

Though the Purusha is free of actions and modifications, yet because its reflection falls upon the mind just as the sun's reflection falls upon a mirror, the mind takes the form of the Purusha and appears to be conscious (*Y.S.*, IV.22).

Purusha's association with Prakrti

Purusha is the detached seer and is associated with Prakrti (seen) in order to gain the knowledge of the Self with the help of Prakrti and to gain the knowledge of the powers of the both (*Y.S.*, II. 23).

When Purusha is identified with the modes of the mind, it continues to experience pleasure and pain through numerous embodiments. But when it attains the highest form of Samadhi, Purusha discovers its mastery over Prakrti and its independence from it. Until this discovery is complete, the contact between the Purusha and Prakrti is maintained. What is the cause of this association?

The cause of the association between Purusha and Prakrti is **ignorance** (*Y.S.*, II.24). Once Purusha realizes its own true nature, it laughs at its ignorance. Ignorance is the latent subconscious impression of wrong knowledge. The intellect laden with such latency does not develop into the illuminating knowledge. When that knowledge arises, ignorance disappears.

Liberation

Once ignorance disappears, the association with Prakrti ceases and the Purusha attains Liberation and rests in its own true nature (*Y.S.*, II.25). Purusha is always independent. But it appears to be bound by Prakrti, because of ignorance. The cause of bondage or liberation, pleasure or pain is in our mind. If you, by oversight, put your finger, you feel the pain, but when you divert your mind to something more important, you do not feel the pain. Everything depends upon the mind. If you think you are bound, you are bound. If you think you are liberated, you are liberated. As you think so you become. When you transcend the mind, you become free from all troubles.

When does ignorance disappear and the Yogi attain final Liberation? When the Yogi attains uninterrupted discriminative enlightenment (*Vivekakhyati*) at the advanced stage of his practice, the ignorance is removed and he ultimately attains Liberation (Y.S., II.26). This discrimination does not mean discrimination between two ordinary materials like salt and sugar. The knowledge of distinction between intellect and Purusha is discriminative knowledge. This knowledge arises first from listening to the scriptures and becomes firmer and clearer through reasoned contemplation. It goes on developing gradually as a Yogi persistently practices Yoga. When he reaches the supreme stage of dispassion through Samadhi, his discriminative enlightenment shines. This enlightenment removes the ignorance, which is the cause for Purusha's association with Prakrti. That ever-shining discriminative illumination is the means of liberation and for cessation of misery (*Y.S.,* II.26). (*See* 3.11 Liberation, below) "The moment you understand yourself as the true Self," says Swami Satchidananda, "you find such peace and bliss, that the impressions of the petty enjoyments you experienced before become as ordinary specks of light in front of the brilliant sun. You lose all interest in them permanently. That is the highest non-attachment."

Purusha's Attainment

Next Patanjali enumerates Purusha's sevenfold attainment. This is called *saptadha bhumi* or seven planes of understanding arising from discriminative discernment (enlightenment). The enlightened Yogi's mind is free from ignorance, distractions and impurities such as greed and jealousy. He gets seven stages of insight:

1. All that has to be known is known. There is nothing more to be known. Until now there was a thirst for knowledge. He was searching for it here and there. At last he has perceived that knowledge, the Truth within himself. He has known that Truth, knowing which everything else is known.
2. He realizes that the cause for pain or misery no longer exists for him. There is no more association with Prakrti. There is nothing further to be avoided.

3. Having attained the highest goal of Liberation the Yogi gets an insight that there is nothing more to be attained.
4. He perceives that all that was to be done was done. The stage of practice has ended. There is nothing more to be done. These four insights are freedom from action. The remaining three types refer to freedom from Prakrti.
5. The liberated Yogi's mind and intellect have fulfilled their purpose. They have given both enjoyment and release to the soul and are no longer needed. The Yogi perceives the cessation of the role of the mind.
6. He perceives that *gunas*, like big stones fallen from the mountain peak never to return, have gone away to merge in their primary cause.
7. Having freed from the play of the *gunas* and the modifications of the mind, the Yogi realizes his true nature and established in the Self, which exists alone. He is established in eternal bliss and peace (*Y.S.*, II.27). He now knows that his Self is pure and perfect and does not require anything else. He does not require anyone else to make him happy, for he is happiness itself.

PRAKRTI

The Nature of Prakrti

Prakrti (primordial Nature) is both the material and the efficient cause of the universe. It is composed of three *gunas—sattva*, (purity), *rajas* (activity) and *tamas* (inertia). It is active and ever-changing, but unintelligent and unconscious. It evolves the world. The *gunas* pass through the phases of equilibrium and imbalance. As long as they maintain their balance, Prakrti remains unmanifested and the universe exists only in a potential state. When the balance is disturbed a re-creation of the universe begins.

Prakrti consists of the gross and subtle elements, and organs controlled by the three *gunas*. The subtle elements consist of sound, touch, color, taste and smell. The gross elements are ether, air, fire, water and earth. The organs include the intellect, mind, senses and the body. The five senses of perception consist of hearing, seeing, tasting, smelling and touching and the five sense of action are verbalization, apprehension, locomotion, excretion and procreation. The eyes, tongue, nostrils and skin are the corresponding physical organs of sense of perception. The mouth, arms, legs, and the organs of excretion and procreation correspond to the senses of action. All these change constantly (*See* 1.4 Samkhya Philosophy and Patanjali's Yoga Philosophy, above)

The Purpose of Prakrti

Prakrti exists for the purpose of giving experience of pleasure and pain and liberation to the Purusha. It does not exist for itself. It is just an **object of the Purusha**. (*Y.S.*, II.18). Prakriti or Nature is here to give us experience and ultimately to liberate us from bondage. Only after getting kicks and burns we get tired of the worldly life and seek freedom from the entanglements, which are like the life of the silk moth, as pointed out by Swami Satchidananda in his commentary on *Yoga Sutras*. The **silk moth story** narrated by him is as follows:

> When the moth is just a day old, it is the size of a hair. Within thirty days, each worm is thicker than a thumb and over three inches long. They grow so big within such a short time because they do nothing but eating mulberry leaves.

The first day hundred worms can feed on a single leaf. The second day a basketful of leaves is needed. The third day, a cartload. The fourth day, a truckload. Day and night they consume the leaves. After thirty or forty days they are so tired they can no longer eat. Then they sleep, as anyone who overeats does. Sleeping on full stomach, the worms roll, and while they roll a juicy type of saliva comes out of their mouths. All that the worms ate comes out as a stream of thick paste, which forms the silk thread. While the worms rotate they become bound up in the thread – the silk cocoon. When all the thread has come out, the warms go into a deep sleep wherein they know nothing.

Finally, they awake to find themselves caught in the tight cage created by their own saliva. "What is this?" the worms think. "Where am I? How did this happen?" Then they remember, "We ate and ate. We enjoyed everything we could, without exception. We over-indulged and became completely exhausted, then totally unconscious. We rolled around and around, binding ourselves in this cocoon. What a terrible thing! We should have at least shared what we had with others. We were completely selfish. We are paying for our mistakes now. Well we repent for all our sins."

The worms repent, pray and fast. In their deep meditation they resolve all their unconscious impressions and decide not to live a selfish life again; at this decision, two wings appear on either side of each worm—one named *viveka* (discrimination), the other *vairagya* (dispassion). These are combined with a sharp, clear intellect, which turns into a sharp nose to pierce open the cocoon. With that, the worms—now silk moths—slip out and fly up high with their fantastically colored wings and look back to see their discarded prisons, "We are leaving and we'll never come back to you again."

There is a beautiful lesson in this story. Like the silk worms, which selfishly ate and ate and got tired and became bound in their own cocoon, we also lead a selfish life, get engaged in restless activities to satisfy our endless desires, enjoying all kinds of sensual pleasures and receive all kind of kicks of worldly problems and sufferings. We bind ourselves by the karmic effects of our selfish actions and sins. When we are tired of all these entanglements we awake and arise from this ignorant muddle and develop discrimination and dispassion like the two wings of the silk moths and seek liberation from the bondage of worldliness by practicing Yoga.

Once Liberation is attained thus by a Purusha, Prakrti ceases to be seen by it. Then it appears to have been destroyed. But it is not destroyed. Why is it not destroyed? Although Prakrti ceases to exist for the Purusha who has attained Liberation, it continues to exist for others who have not yet attained Liberation, as it is common to them. They need the sustenance of Prakrti for their spiritual evolution and ultimate Liberation (*Y.S.*, II.21).

The Samkhya Philosophy says: Prakrti will always remain as it is now. Its total destruction is not possible, because living beings to attain liberation are innumerable. So the association between Prakrti and Purusha is endless.

Questions

1. What is the nature of Purusha?
2. Why does it appear to experience pleasure and pain?
3. Is the Purusha the only seer? How?
4. Why is the Purusha associated with Prakrti?
5. How does a Purusha attain liberation?
6. Explain the seven stages of insight that the Purusha gets.
7. What is the nature of Prakrti and what is its purpose?
8. Does the Prakrti cease to exist? Why or why not?

2.9: God (Isvara)

Introduction

In the Samkhya Philosophy, on which Patanjali's Yoga Philosophy is based, there is no reference to God. But Patanjali accepts God. According to the Vedas God is the creator and preserver of the universe. But Patanjali does not associate with God the idea of creating and preserving the universe.

Patanjali's concept of God

Patanjali considers God as a special Purusha untouched by afflictions, actions, their effects and latent impressions of actions (*Y.S.*, I. 24).

Afflictions (*klesas*) are caused by ignorance, egoism, attachment, aversion, and clinging to life (*see* 2.6 Afflictions: Causes and Remedy, below).

Karmas are good, bad and mixed actions (*see* 2.7 The Doctrine of Karma, above).

Effects of actions: Actions performed with selfish desire and attachment to their fruits give rise to three types of effects – birth in a particular kind, tenure of life, and experience of pleasure and pain.

Impressions of actions exist in subtle form in the subconscious mind, and they fructify when favorable conditions develop.

An individual soul is affected by these limitations. By meditating upon God and surrendering to Him, a Yogi liberates his soul from the bondage of karmas and fetters of Prakrti.

While individual souls are bound by karmas, God is always free from bondage and limitations. When the embodied soul (*jiva*) destroys ignorance by knowledge, it communes with God and discovers its identity with the Absolute. God's pre-eminence knows no comparison, and is unsurpassable by anybody.

Omniscience

In God there is unexcelled omniscience. In Him exists infinite knowledge; in others knowledge is only a germ. God is the very embodiment of Consciousness. He knows everything in respect of the past, present and future (*Y.S.*, I.25). When a Yogi begins to meditate on God the seed of knowledge begins to unfold in him.

God is not something that can be talked about, it must be experienced. If a drop of water wants to know the depth of the ocean, it must become the ocean. So also if one wants to know God, one must become God. That is why Narada, the divine sage, has avoided defining God anywhere in his *Bhakti Sutras*, because to define is to limit God. God , the ultimate Reality, is to be experienced. In the words of Sri Ramakrishna, God- incarnate, "When one attains Samadhi, then alone comes the knowledge of Brahman. In that realization all thoughts cease and one becomes perfectly silent. There is no power of speech left by which to express Brahman." The *Brahadaranyaka Upanishad* declares,

> "This Immutable is never seen, but is the Seer; it is unheard, but is the Hearer; it is unknown, but is the Knower. There is no other seer but This, no other hearer but This, no other thinker but This, no other knower but This" (*Y.S.*, III.8.11).

The Vedanta describes Brahman as *Satchidananda*, which means Existence-Knowledge-Bliss Absolute. "Brahman exists as the pure Consciousness within and beyond all finite forms of consciousness, and as Shakti, the primordial Energy. Brahman and Shakti are identical like the fire and its power to burn," says Sri Ramakrishna. Thiruvalluvar in his sacred *Thirukkural* describes God as pure Intelligence or Consciousness (*valarivan*), one beyond likes and dislikes (*Ventuthal vayntaamai ilaan*), one who is free from desires (*Porivayil aintavittaan*), Incomparable One (*Thanakkuvamai illathaan*) and Ocean of Virtue (*Aravaazhi*) (Chapter 1. *Kadavul Vaalthu*)

Teacher of teachers

God is the teacher of even the most ancient teachers. Although all knowledge is with you, a teacher is still necessary to remove the veil of ignorance that envelops the knowledge within you. Your teacher must have been learned from his teacher. That teacher must have been leaned from another teacher. Thus there is a chain of teachers. But there should be a primary teacher. God is the first teacher, but to Him, time as a limiting factor is not applicable (*Y.S.*, I.26). The *Bhagavad Gita* and the *Upanishads* declare that God taught even Brahma, the creator before the creation of the universe.

OM

God must have a name, but it should not be a limited one. The name to denote the infinite God must give an unlimited idea and vibration. Such a name cannot be anything but OM (*Y.S.*, I.27). The *Mandukya Upanishad* declares,

> "AUM, this syllable is all this. All that is the past, the present and the future, all this is only the syllable AUM. And whatever else there is beyond the threefold time, that too is only the syllable AUM" (1).
>
> "All this is verily Brahman. This Self is Brahman" (2).

The syllable AUM, which is the symbol of Brahman (God), stands for the manifested world, the past, the present and the future, as well as the unmanifested Absolute.

"In the beginning was Brahman," says *Rick Veda*, "with whom was the word and the word was truly the supreme Brahman." There is a similar statement in the Gospel: "In the beginning was the word, and the word was God." OM (AUM) was that word. It is the root of all mantras (sacred syllables). It is the most basic, the most natural of all sounds.

This sound OM is pronounced as AUM. The first letter A is the root sound pronounced by simply opening the mouth without touching the tongue or palate. Then as the sound comes forward between the tongue and the palate up to the lips U or "OO" is pronounced. Then the closing lips produces M, "Mmmm" or "hum."

That hum is called *Pranava*, because it is connected with *prana* (life force). *Prana* is the basic vibration that always exists in manifested and unmanifested condition. Even without your repetition, the basic sound is always vibrating in you. It is the seed of all other sounds. Hence OM represents God in the fullest sense. Lord Krishna proclaims in the *Bhagavad Gita*, "I am *Pranava* (OM) in the *Vedas*" (7.8).

OM was not invented by anybody. God Himself manifested as OM. "Amen" or "Ameen," which the Christians, Muslims and Jews say is a variation of this OM.

Repeating OM

OM is a basic mantra to be repeated. The word "*mantra*" means "that which keeps the mind steady and produces the proper effect." Its repetition is called Japa. Mental repetition is most effective. The repetition of OM and contemplation on its meaning bring one-pointed focus to the mind of a Yogi. It is a powerful technique and, at the same time, the easiest, simplest and the best. When the mind is fixed on OM, no other thought arises in the mind. Therefore, almost every religion advocates the repetition of God's name. In fact all the prophets, sages and saints experienced God and understood Its greatness, glory and power by repeating OM. It leads to communion with God. Along with each repetition the mind should be directed to God by thinking of His nature. The repetition of OM purifies the mind. The *Bhagavad Gita* says,

> "...Among words I am the syllable OM; among sacrifices (*Yajna*) I am the silent repetition . . ." (10.25).

When OM and its import come naturally to the mind, self-surrender to God is well established. This is beautifully described in the *Mundaka Upanishad*:

> "The syllable OM is the bow. One's self indeed is the arrow.

Brahman (God) is its target. It is to be hit with a one-pointed mind. Thus one becomes united with Brahman as the arrow (becomes one with the target)" (II. 2.4).

On attaining perfection in both the repetition of OM and the meditation on OM, the Supreme Self shines forth (*Y.S.*, I.28).

Devotion to God

By the practice of repeating OM all obstacles to meditation such as sickness, dullness, etc. disappear and the Yogi begins to ascend the heights of Samadhi. His mind is purified, and the knowledge of inner Self dawns. He realizes that the immortal Self in him is different from his body, mind, senses, intellect and ego. Just as God is pure, isolated, changeless, and free from afflictions, so the Yogi's soul becomes pure and free from afflictions.

The repetition of OM strengthens Yogi's devotion to God. By intense devotion to God he attains Samadhi and realizes his inner Self, transcending all his limitation (*Y.S.*, I.29). The Yogi gets in tune with the Cosmic power. By that tuning he feels the force in himself, imbibes all the Divine qualities, gets the Cosmic vision, transcends all his limitations, and finally becomes that transcendental Reality. Normally the mind and body limit us, but by holding something infinite we slowly raise ourselves from the finite things that bind us and transcend them. Through that all obstacles to meditation disappear.

Self-surrender to God

Saint Patanjali declares that **self-surrender to God is an alternate means for attaining the highest Samadhi** (*Y.S.*, I.23). Yogis attain the highest Samadhi at the final stage of their meditation practice. *Bhaktas* (Devotees) attain the highest Samadhi by total dedication or self-surrender to God.

Self-surrender means offering our body, mind, ego, senses and the soul at the feet of God.. It is complete dedication of us to the Supreme Self. Then the self (ego) within us is dead and God is enthroned in its stead, and there is no longer any "I, me, my or mine." Self-surrender is a state in which the thought of God runs in the mind in a ceaseless and continuous stream. It involves a total change in our attitude towards us, towards the world, and towards God. Is there anything that we can claim as "ours?" There is only one thing. That is God. Everything else – human beings, other living beings, objects of the world, and the entire universe – all is God's manifestation.

Perfect self-surrender is the culmination of spiritual life. It is the final step in the path of devotion. God is realized. As a river merges into the ocean, a seeker merges his small will and ego into the Supreme Power, which is the doer of all things. "It is all His Name, it is all His Form, It is all His Deed, and it is all for good," says Swami Satchidananda. Lord Krishna says in the *Bhagavad Gita*,

> "Among all the Yogis, I consider him to be the most devout who worships Me with full faith, and with his inner self merged in Me" (6.47).

Questions

1. Explain Patanjali's concept of God.
2. How does Patanjali describe God's unexcelled nature?
3. Discuss the glory of OM. How is it the most appropriate name to denote God?
4. Explain the importance of repeating OM.
5. What is self-surrender? What is its significance?

SECTION III

Patanjali's Yoga System

3.1: Kriya Yoga

Introduction

Mind is polluted by endless desires, delusion, lust, greed, anger, anxiety and other impurities. The impure mind, like a dirty mirror, cannot reflect the clear image of the Self or Atman or Divinity within.

The root cause for the impurities and the consequent sufferings is ignorance, i.e., unawareness of one's own reality. Out of ignorance of our real nature we wrongly identify ourselves with our bodies, minds and senses and indulge in selfish activities and commit vicious deeds. Consequently we reap the reactions of our actions in the form of miseries. Our ignorance, therefore, is the root cause of miseries.

Ignorance leads to egoism. From egoism arise attachment to desirable things and aversion to undesirable things. From one's attachment to objects and people and one's sense of egoism come the deep attachment to life and aversion to death.

All these pollute the mind, lead to karmic bondage and cause pain, affliction or misery.

One has to purify the mind in order to attain success in the practice of Yoga and realize one's own reality.

Kriya Yoga

Patanjali suggests an **action program** for purifying the mind. It is called Kriya-Yoga. It is **not a separate kind of Yoga**. It is just a **purifying practice**. Kriya Yoga means actions performed for purifying the mind. The actions primarily consist of **austerity, study of scriptures** and **surrender to God** (*Y.S.*, II.1).

The Purpose of Kriya Yoga

The purpose of Kriya-Yoga is to **attenuate** or reduce the **causes of afflictions** (*klesas*), viz., ignorance, egoism, attachment, aversion and clinging to life in order to **attain success in Samadhi** (*Y.S.*, II.2).

The practice of Kriya Yoga reduces the causes of afflictions and thus purifies the mind. The thinning out of causes promotes success in Samadhi, which leads to discriminative discernment or knowledge. The fire of this knowledge destroys ignorance and eventually completes the process of destruction of other causes of afflictions. The Yogi, then, is able to reach the highest Samadhi and ultimately attains Self-realization (*See* 3.10 Samadhi, and 3.11 Liberation, below).

The actions (*kriyas*) that constitute Kriya Yoga are described below.

Austerity (*Tapas*)

Tapas means "to burn." It is meant for "burning out" impurities. Anything burnt out is purified. The more you fire the gold, the more pure it becomes. One form of austerity is **accepting all pain** that comes to us. Such acceptance makes the mind serene, steady and strong, for it is hard to accept pain caused by others without returning it.

Tapas also implies **self-discipline**. The mind is like a wild horse tied to a chariot. Imagine the body is a chariot, the intellect is the charioteer, the mind is the reins, and the horses are the senses. The self is the passenger. If the horses are allowed to gallop without reins and charioteer, the journey will not be safe. If the senses are controlled, the passenger will reach the end of the journey (*Katha Upanishad*, I.3.3-9).

Simple living, control of senses, sexual purity, regulated fasting, observance of silence for a day every week, cultivation of moral qualities, chanting the name of God and selfless service – all these constitute *tapas*. Through *tapas* the mind is disciplined and purified and willpower is developed.

In the name of *tapas* some people indulge in all sorts of self-torture like lying on a bed of nails. Lord Krishna, in the *Bhagavad Gita*, says that people who torture their bodies are demons because they disturb the pure Self that dwells within their bodies. He prescribes moderation:

> "Yoga is not possible for a person who eats too much, nor for him who does not eat at all; nor for him who sleeps too much, nor for him who is always awake" (6.16).

Self-discipline is absolutely necessary for spiritual progress. It will not obstruct one's normal life. Rather it will make one's life meaningful and peaceful.

Self-study (*Svadhyaya*)

This means **study of scriptures** pertaining to the Knowledge of Self and Liberation. Scriptures like the *Upanishads*, the *Bhagavad Gita, Thirumantiram, Thirukkural, Atma bodha, Yoga Sutras, Bible* and others that will elevate our minds and strengthen our conviction about our true nature should be studied and understood. Mere learning is not enough. What is learnt should be put into practice.

Self-study also means self-analysis or looking at oneself with greater objectivity, analyzing how different things affect one in everyday life. This helps one to correct one's false notions, wrong attitudes and inner conflicts.

Self-surrender to God (*Ishwara-pranidhana*)

This means dedication of everything – our thoughts, words, deeds, our life and soul to God. Nothing is ours. We should pray, "I am Thine. All is Thine. Thy will be done." Total surrender to God will remove the veil of egoism and selfishness, and it will bring permanent peace and joy. Our entire life will become an offering at the Lotus Feet of the Lord. We should perform selfless service to humanity without any personal motive. Service to other beings is service to God. That is a form of worship. The ancient great saint Thirumular of Southern India beautifully describes how service to humanity is service to the Lord, in his sacred scripture *Tirumantiram*:

Padamada koyil Bhagavarkon treyil
Nadamada koyil nambark ankaka
Nadamada koyil nambarkon treyil
Padamada koyil Bhagavarka thame (1857).
Offering given to the Lord in the towered temple
Reaches not the noble walking temples.
Offering given to the noble walking temples
Reaches surely the Lord in the towered temple.

The offerings of such things as fruits, milk, honey, cooked food, etc. made to the Lord in the temple do not reach the Lord's noble devotees, who are walking temples. But the things offered to the devotees surely reach the Lord in the temple. Thus service to humanity and also service to other living creatures is service to God.

Total surrender to God and inward renunciation of all worldly and religious concerns does not mean abandonment of actions beneficial to the world. As an instrument in Divine hands the devotee continues to perform actions, giving up their fruits to the Lord. (*Narada Bhakti Sutras*, 62). Lord Krishna, in the *Bhagavad Gita*, says:

> "Whatever thou do, whatever thou eat, whatever thou offer in sacrifice, whatever thou give, whatever thou practice as austerity, O Arjuna, do it as an offering unto Me." (9.27)
> "Thus shall thou be freed from the bonds of actions yielding good and bad fruits; with the mind steadfast in the Yoga of renunciation, and liberated, thou shall come unto Me. (9.28)

Consecrate all actions to the Lord. Perform actions without any sense of doership and without any attachment to actions and their fruits. Then you are freed from the bondage of Karma. You attain freedom in action. Your actions thus become selfless service and worship of the Lord. You live for the Lord only. You work for the Lord only. When actions are dedicated to the Lord, there is no rebirth for you.

In this path of self-surrender, all actions, all results and all rewards go the Lord. There is no separate life for you. Just as the river joins the ocean and abandons its name and form, so also your soul joins the Supreme soul, giving up your name and form. You attain liberation, while living, and when your body falls you become one with the Lord. Take, for instance, the lives of great teachers like Krishna, Buddha, Jesus Christ, Sri Ramakrishna and others. They had realized God; they became One with God. There was nothing more to be achieved by them. Yet they continued to work for the benefit of the society without any personal or selfish motive.

Of course self-surrender is a gradual process. We have to consciously and sincerely practice it with intense devotion without a break in order to attain success in it.

There are various *kriyas* or actions of purification in addition to the above three recommended by Patanjali, but these three constitute the first step in the path and provide the foundation for all other practices. (*See* the last Paragraph in 2.9 God (Isvara), above).

The Outcome of Kriya Yoga

The benefits of Kriya-Yoga are as follows:

First, the effective practice of Kriya Yoga **weakens** or attenuates considerably the **causes for afflictions**—ignorance, egoism, attachment, aversion and clinging to life.. When these are considerably weakened, they become incapable of causing the modifications of the mind. Then the Yogi is able to make progress in his practice of Yoga. When he attains discriminative knowledge, the causes for afflictions are burnt up in the fire of knowledge. They, then, cannot cause modifications of the mind.

Second, the practice of Kriya-Yoga **purifies the mind**. The purified mind becomes fit for concentration...The Yogi is able to sustain meditation and makes progress toward Samadhi. The Yogi is able to extend the duration of concentration on the object of meditation, and gradually **advances toward the highest Samadhi**. Then discriminative knowledge dawns, and the Yogi frees himself from the Prakrti (*Y.S.*, II.2).

Questions

1. How does the mind is polluted? What is the cause for the impurities of the mind?
2. What is Kriya Yoga? What is its purpose? How is it accomplished?
3. What are the actions that constitute Kriya Yoga? Describe them.

3.2: Raja Yoga: An Outline

Introduction

The system of Yoga described by Patanjali in the *Yoga Sutras* is ***Raja Yoga***, the royal path. Raja Yoga is one of the major kinds of Yoga. It is as important as the other kinds of Yoga such as Bhakti Yoga, Karma Yoga, Jnana Yoga and Japa Yoga. Raja Yoga encompasses the teachings from all the different paths. It is a comprehensive system. It is the **most systematic method of attaining Superconscious state** through Meditation.

Raja Yoga is Yoga of Meditation and also includes all ethical and moral observances, *Asanas* and *Pranayama* as its prerequisites. It is a comprehensive **scientific discipline**. It can be practiced by people of different faiths, backgrounds and temperaments. It is best suited to those who have a more experimental mystical nature. It envisages all aspects of human life: physical, mental and spiritual. The prescribed practices of Raja Yoga can be scientifically verified by anyone who accepts its methods. It is ideally suited to modern times whose hallmark is verifiability or testability for accepting any hypothesis.

Eight-step Path

Patanjali divides the path of Raja Yoga into eight limbs or stages (*Yoga Sutras*, II.29). Therefore it is also called *Astanga Yoga* (eight-limbed path). The eight stages or steps constitute a systematic path of regulation and discipline in a logical and natural sequence from the gross physical body to the subtler senses, and to the subtlest levels of consciousness. The **eight stages** or steps are:

Preparatory Practices (*Bahiranga Yoga*)

1. *Yama* (Self-restraints)
2. *Niyama* (Observances)
3. *Asana* (Posture)
4. *Pranayama* (Regulation of breath), and
5. *Pratyahara* (Withdrawal of senses)

Higher Stages (*Antaranga Yoga*)

6. *Dharana* (Concentration)
7. *Dhyana* (Meditation)
8. *Samadhi* (Absorption or Superconscious State).

The three-fold higher process of Concentration, Meditation and *Samadhi*, by which the true nature of an object is known is called *Samyama* (*Yoga Sutras*, III.4).

The above eight-steps will be described briefly here, and in succeeding Chapters they will be discussed separately in detail.

The **first five stages**, though not directly connected with Meditation, are of the **utmost importance**, for they **prepare an aspirant's mind and body for the higher stages.** Without some practice of the first five stages, most people may not be successful in Meditation. The first five stages constitute the means for overcoming the obstacles to Meditation. They will mold a person's character in the way necessary for spiritual progress. Of course some people will be able to meditate without even knowing the existence of those stages, but they are the few lucky people who have no mental or physical obstacles, and who from birth have been inclined to looking inward and toward the meditative way of life.

It becomes easier to practice the higher three stages when the preliminary steps have been practiced to a reasonable degree of perfection. The reason for this is that most of us are totally unable to concentrate and meditate because of the continuous wandering of the mind. Only a person with a tranquil mind can be able to meditate.

The types of **disturbances** that prevent Concentration and Meditation are:

(a) **Emotional disturbances** arising out of mental conflicts and imperfections. These are eliminated or at least reduced by practicing the *Yamas* and *Niyamas* (stages 1 and 2).

(b) **Physical discomforts**, such as pain, illness and uncomfortable posture. These are removed by practicing *Asanas* (stage 3).

(c) **Irregularities in the *pranic* flow** in the body, which cause disturbances. The technique of *Pranayama* (stage 4) corrects such irregularities.

(d) **Outside distractions** such as sound, which cause mental disturbance. How is it possible to perform the inner techniques when our mind is absorbed and continually distracted by the outside environment? *Pratyahara* (stage 5) eliminates this source of disturbance by disconnecting the association of the sense organs, eyes, ears, etc. with external happenings. The outer occurrences are still there, of course, but the sense organs no longer send messages to the mind or the mind does not become aware of them.

One can now realize how important are the first five stages in order to practice successfully the higher stages. Without cultivating them, the practitioner will be unable to reach the higher stages.

Yama and *Niyama*

These first two steps consist of ten Commandments of Yoga. The *five Yamas* (Restraints) are non-violence, truthfulness, non-stealing, continence, and non-possessiveness (*Y.S.*, II.30). Their practice leads to behavioral modifications and replacement of negative qualities by virtues. The five *Niyamas* (observances) are cleanliness or purity, contentment, austerity (*tapas*), self-study and surrender to God (*Y.S.*, II.32). The *Yamas* and *Niyamas* involve moral training without which no practice of Yoga will be successful.

The beginner should not be discouraged by the immensity of these first two steps of Raja Yoga. He is not expected to perfect the practice of *Yama*s and *Niyama*s before proceeding further, but he should try to practice them as conscientiously as he can. With persistent effort he will eventually be able to perfect them. In the West in Yoga centers, in general, *Asanas* and breathing exercises only are taught. The *Yama*s and *Niyama*s are usually neglected, because of the difficulties they entail and the changes of life style that are necessary to practice them. It is true that *Asanas* and breathing exercises ensure physical health, but their full benefits can be realized only when the mind is purified and becomes steady and tranquil by practicing the *Yama*s and *Niyama*s.

Asanas

The third step in Raja Yoga is *Asanas* or Postures. There are **two types: Meditative Postures** and ***Asanas* for physical wellness**. A steady Meditative Posture leads gradually to a stable mind. A posture suitable for Meditation should be comfortable and stable, ensuring that the head, neck and chest be erect and in a straight line.

The second kind of postures is practiced to perfect the body, making it supple and free from disease. These postures control specific muscles and nerves and hence have therapeutic benefits.

Pranayama

This is the fourth step of Raja Yoga. It means **regulation of *prana***, the vital energy that sustains the body and mind. The grossest manifestation of *prana* is the breath, so *Pranayama* is also called the science of breathing. Regulation of breath leads to regulation of the mind. It also purifies and strengthens the nervous system.

Pratyahara

It means **withdrawal** and control **of the senses**. The mind comes into contact with the objects of the world through the five senses of seeing, hearing, touching, tasting and smelling. These sensual experiences set the mind to wandering. The aspirant should therefore acquire the ability to voluntarily draw the senses inward and thus isolate himself from the distractions of the world outside.

Dharana

Dharana or Concentration is the sixth step in Raja Yoga. In Concentration the dissipated powers of the mind are gathered together and directed towards the object of Concentration through continuous voluntary attention. Through Concentration the diffused mind is **focused** and hence made more powerful and penetrative.

Dhyana

Prolonged, unbroken Concentration leads to *Dhyana* or state of Meditation, which is the seventh step in Raja Yoga. Concentration makes the mind one-pointed and steady. Meditation expands the focused mind to the Superconscious state by piercing through its conscious and subconscious levels. All methods of Yoga prepare one to reach the stage of Meditation, for only through Meditation can one reach the level of the Superconscious mind and hence attain perfection. Meditation alone can take one to this blissful state of mind. It is possible to attain that state only through persistent and intensive practice. Modern education is superficial, one-sided. It does not help one to know, develop and control one's internal states. Meditation alone can help us to do this. We then become aware of our latent powers and become more creative and dynamic.

Samadhi

Prolonged and deep Meditation leads to the last stage of Raja Yoga, the state of *Samadhi* or Superconscious state. In this state one becomes **One with the Divine Self** (*Paramatman*) and transcends all imperfections and limitations.

The physical sciences are based on sense perceptions that are interpreted by the rational faculty. Sense perceptions and rationality, however, are limited by time, space and causation. They do not give us a complete understanding of the various forces that shape us and the universe. Yoga science is the soul of all sciences. It solves the basic problems of life. It is definite and systematic and it leads one to the highest source of knowledge, intuitive knowledge that reveals our true nature..

The teachings of Yoga have been handed down to us from the ancient Yogis of India. A sincere aspirant who follows these teachings can reach the ultimate state and experience the infinite consciousness. He can discover for himself his true nature, the Atman or Brahman within.

Questions

1. What is Raja Yoga? Briefly describe its eight stages.
2. Why are the first five preparatory stages considered important?
3. What types of disturbances do prevent concentration? How are they eliminated?

3.3: Obstacles to Yoga

Introduction

The practice of Yoga requires peace of mind. If the mind is not calm it is not possible to concentrate and sustain it. Various obstacles disturb the peace of mind and affect the practice of Yoga. Patanjali enumerates them in his *Yoga Sutras:*

Sickness, dullness, doubt, negligence, sloth, sensuality, false perception, failure to reach firm ground, and inability to stay in the stage of concentration attained – these distractions to the mind are obstacles to Yoga (*Y.S.,* I.30).

Grief, dejection, trembling of the body and irregular breathing are accompaniments to the mental distractions (*Y.S.,* I.31).

Sickness

Sickness is caused by an imbalance in the humors or in the secretions or in the circulation of blood. Irregular and improper dietary habits, overeating, starvation, virus attack, accidents, etc. cause sickness. Disease affects the mind. It causes aches and pain, and also dullness and laziness. It also causes discomfort When the body is sick concentration is not possible. One can maintain health by taking fat-free vegetarian diet and leading a well-regulated and disciplined life. One must also practice Hatha Yoga asanas and do simple exercises like walking.

Dullness

This is the inactivity of the mind It affects the inclination to practice concentration. A dull mind refuses to penetrate into a thing and doubts everything.

Doubt

Doubt arises due to failure to contemplate and lack of conviction about the scriptural maxims, and the truth of the science of Yoga, and so on. If there is doubt, firmness is not possible, until some psychic experiences come. These glimpses strengthen the mind and give faith and encouragement. By listening to spiritual discourses, associating with successful Yogis, and reflecting over scriptural messages and words of sages and saints, one can get rid of doubts.

Negligence

Negligence is lack of sustained and dedicated practice of Yoga. It also arises due to lack of genuine interest in Yoga. Pre-occupation with too many worldly affairs also will lead to negligence in the practice of Yoga.

Sloth

Sloth is a quality of *tamas*. It is the inactivity of body and mind due to inertia or laziness. It tempts one to relax, to neglect duties and to resist self-discipline. In sloth the mind remains torpid. Moderation in diet, alertness and enthusiasm can conquer sloth.

Sensuality

Lust and greed lead to intense desire for objects of senses and craving for sensual experiences. They lead to restless activities for amassing wealth by fair or vicious methods. Indulgence in sensual pleasures and enjoyments sap the vitality and in due course cause nervous disability and deadly diseases. One who cannot control senses cannot attain any success in Yoga and also cannot develop moral strength. Moderation in diet, work, rest, etc. is essential for Yogic life.

False Perception

This is erroneous perception or false knowledge. It means considering lower stage to be higher stage and vice versa. Listening to spiritual discourses and deep study of scriptures and association with the wise will remove false conception.

Failure to reach firm ground

Failure to attain progress and reach some definite stage in concentration is due to negligence in purifying efforts and irregular practice.

Inability to stay

This means failure to maintain the attained state of concentration. This is due to negligence, lack of sustained interest in Yoga and irregular practice. One should develop firm commitment and determination to practice regularly.

Grief or pain

Pain is of three kinds: (1) pain arising within one's own body or mind due to sickness or accidental hurt, (2) pain inflicted by other creatures like snake and scorpion, and injuries caused by others, loss of property or death of loved ones and (3) pain caused by natural calamities like flood, earthquake, cyclone, etc. Life itself is a continuous effort to overcome grief.

Dejection

Dejection is the agitation of the mind due to non-fulfillment of desires. It also arises when expected things do not happen. It is an outcome of egoism and selfishness. Unless one develops equanimity, one cannot avoid dejection. Self-surrender to God will help to avoid grief.

Unsteadiness of the body

The disturbance of bodily equilibrium makes the body to tremble. An agitated mind causes tension in the nerves, resulting in abnormal restlessness in the body.

Irregular breathing

Irregular, imbalanced and jerky breathing is an obstacle to Yoga. It is caused by mental disturbances and emotional problems. It disturbs the even flow of energy in the body and affects the digestive and other systems. By practicing Pranayama regularly breath can be regulated. Then the mind becomes calm and steady (*See* 3.6 Pranayama, below)

Overcoming Obstacles

By self-surrendering to God, seeking good association, taking right diet, exercise and rest, cultivating positive qualities, and regulating daily life, a Yogi can gradually gain powers with which to resist and overcome the above obstacles.

Sri Swami Sivananda, a great saint and the founder of Divine Life Society and Sivananda Ashram in Rishikesh, Himalayas, India, gives the following general advice to overcome obstacles:

- Assimilate the Divine ideas. Saturate the mind with thoughts of Brahman, Divine glory and sublime spiritual thoughts in order to become established in the Divine consciousness.
- Free yourself from the base thoughts of the mind, and useless imaginations. Just as you render the turbid water pure by adding clearing nut, so also make the turbid mind pure by Divine thoughts.
- Purge the mind of all impurities, then it will become calm.
- Spiritual practice is a life-long process. Obstacles are innumerable in this great voyage. But so long as you hold God as your guide, there is nothing to worry about.
- If you conquer one obstacle, another one is ready to manifest. If you remove greed, anger is waiting to hurl you down. Therefore great patience, perseverance, vigilance and undaunted strength are needed.

The best way to overcome the obstacles, says Patanjali, is to **practice regularly concentration** on a chosen single object of concentration such as OM, chosen deity or mantra given by the Guru (*Y.S.*, I.32). One should not keep changing the object of concentration, but stick to one thing. One-pointed concentration will bring perfect repose to the mind and body.

Maintaining the Peace of Mind

There are **several ways** to maintain the mind peaceful.

1. One of them is our **attitude towards others**. What is the proper attitude? This depends upon the conditions of the person whom we come across. Patanjali gives four keys: **friendliness, compassion, joy** (delight), and **indifference** for four different persons – happy, miserable, virtuous, and the wicked.

We should entertain a spirit of friendliness towards happy people, instead of feeling envious. We should be merciful and compassionate towards those who are in distress. If possible we should extend a helping hand to them.

When we come across a virtuous person, we should feel delighted and appreciate his virtuous qualities. Sometimes we may come across a wicked person. We should be indifferent to him with a spirit of benevolence. We should not try to advice him because wicked people seldom take advice. Rather we should silently pray for their transformation. If we try to advice a wicked person we will only incur his wrath and lose our peace. Here is a **short story** from *Pancha Tantra*, which was narrated by Swami Satchidananda in his commentary to illustrate this point.

One rainy day, a monkey was sitting on a tree branch getting completely drenched. Right opposite on another branch of the same tree there was a small sparrow sitting in its hanging nest. Normally a sparrow builds its nest on the edge of a branch so that it can hang down and swing around gently in the breeze. It has a nice cabin inside with an upper chamber, reception room, a bedroom down below and even a delivery room if it is going to give birth to little ones. A sparrow's net is admirable.

So it was warm and cozy inside its nest and it just peeped out and saw a monkey well drenched. The sparrow said, "Oh, my dear friend, I am so small; I don't even have hands like you, only a small beak. But with only that I built a nice house expecting this rainy day. Even if the rain continues for several days, I will be warm inside. It is said you are the forefather of the human beings, so why don't you use your brain? Build a nice, small hut somewhere to protect yourself during the rain."

The monkey was terribly angry, "Oh, little devil! How dare you are to advice me? Because you are warm and cozy in your nest you are teasing me. Wait, you will see where you are!" The monkey started to tear the nest to pieces, and the poor bird had to fly out and get drenched like the monkey.

Sometimes we come across wicked persons like that monkey, and if you advice them, they take it as an insult. If you sense even a little of that tendency in somebody, stay away. He or she will have to learn by experience. By giving advice to such persons you will only lose your peace of mind.

Patanjali groups all people in these four categories: the happy, the virtuous, the unhappy and the wicked. So let us have these four attitudes: friendliness, gladness, compassion and indifference. These four keys should always be with us. If we use the right key to the right person, we will retain our peace. Nothing, then, will upset us. Our goal is to keep a serene mind. (*Y.S.*, I.33).

But in our daily life often we are unable to keep our minds in this way, and consequently face difficulties and disturb our minds. For instance if someone does evil to us, instantly we react with evil. Every negative reaction like provocation, anger, etc. brings out a chain reaction and we lose our mental balance and power. By controlling the thoughts of reaction and revenge we can prevent mental disturbance and conserve good energy. The great saintly poet (of Tamil Nadu, Southern India) Thiruvalluvar in his sacred *Thirukkural* advocates not only to forgive evil done to us by a wrongdoer but also to forget it. He goes further and emphasizes doing good to him in return:

Innaa ceythaarkkum inniyavay ceyaakkaal
Enna payathatho saalpu (Kural 987).

This means: Of what avail is sublimity, if something good is not returned for evil? Thiruvalluvar upholds the virtue of doing good to one who hurts you:

Innaa ceytaarai orutthal avarnaana
Nannayam ceythu vital (Kural 314).

That is, to punish a wrongdoer is to do good to him and then to forget both the wrong done by him and the good done by you in return. If this attitude is cultivated mind you will ever remain serene and peaceful.

2. Or mental peace is maintained **by regulating the flow of *prana*** through the regulation of the breath (*Y.S.*, I.34). This is called Pranayama. There is a close connection between the mind and *prana* (vital energy or life-force). The great South Indian saint Thirumular says, "Where the mind goes, the *prana* follows." This is seen in our daily life. If your mind is agitated, you will be breathing heavily. If you are reading something deeply, you are hardly breathing. So if you regulate the *prana*, your mind is regulated automatically.

3. Or the **concentration on subtle sense perceptions** makes the mind steady (*Y.S.*, I.35). During the initial practice of Concentration various extraordinary sense perceptions may occur. The Concentration on the tip of the nose will give you a fine smell. Concentration on the tip of the tongue will give you a nice taste, and so on.

4. Or by **concentrating on the Effulgent Light**, you gain stability of the mind (*Y.S.*, I. 36). Imagine luminous lotus in your heart and concentrate on it, you will get a nice experience of brilliant light; the mind is stabilized.

5. Or **by concentrating on a great soul's mind** that has given up all attachment to sense objects, the mind gains stability (*Y.S.*, I.37). Take some saint whom you know to be perfectly non-attached, and think of his heart. Meditate on that heart, it will calm your mind.

6. Or by **concentrating on an elevating experience** you had in a dream or sleep, the mind can gain steadiness (*Y.S.*, I.38).

7. Or by **meditating on anything that you consider elevating**, you gain peace of mind (*Y.S.*, I.39).

Patanjali has suggested a few alternate methods. Yet some may not find any of those techniques appealing to them. Therefore he concludes, saying, "You can concentrate on anything that you consider as elevating."

Questions

1. What are the obstacles to Yoga? Briefly describe each of those?
2. What are the suggestions made by Patanjali to prevent the obstacles?
3. How can the peace of mind be maintained?

3.4: *Yama* and *Niyama*

YAMA

Introduction

Yamas and *Niyamas* are the first two stages or limbs of Raja Yoga. *Yamas* consist of five self-restraints that regulate one's relationship with others. They are: non-hurting (*ahimsa*), truthfulness (*sathya*), non-stealing (*asteya*), celibacy (*brahmacharya*) and non-possessiveness (*aparigraha*) (*Yoga Sutras*, II.30). The *Niyama*s are five observances to be followed. (*See* next subheading, below). These ten commitments constitute the **moral code** of Raja Yoga. They are similar to the Ten Commandments of the Christian and Jewish faiths, and the ten virtues of Buddhism. In fact, there is no religion without these moral codes. All spiritual life is based on these codes.

The *Yama*s and *Niyama*s are closely connected with higher Yoga. They are meant for purifying the mind. There is direct interaction between the mind and the body. Physical health is dependent on mental health. Scientific findings indicate that most diseases are caused by mental and emotional disturbances. It is therefore necessary to cultivate attitudes and values that ensure a steady and peaceful mind. Mental peace and sound physical health are pre-requirements for practicing Yoga. The *Yama*s and *Niyama*s seek to remove all mental and emotional disturbances and thus to make the mind peaceful and serene. That is why the *Yama*s and *Niyama*s have been included as the first two stages of Raja Yoga.

Morality and Yogic Life

What is the place of morality in Yogic life? There are two kinds of Yoga: lower and higher. The lower Yoga is practiced for attaining certain psychic powers. There is a large number of Yogis scattered throughout India, Tibet and other countries, who possess supernormal powers, but in moral standards they are not different from the ordinary people. Some of them are good people. Others are not innocent and can cause injury to those who cross their path. There is a third class of Yogis who are dangerous and misuse their powers for selfish purposes. The lower Yoga is nothing but misapplication of Yogic practices for self-aggrandizement or satisfaction of conceit and egoism.

The higher Yoga expounded in the *Yoga Sutras* is the genuine Yoga that is used for attaining Enlightenment, and freedom from the illusions and limitations of the lower life.

For the path of higher Yoga, morality of a high order is essential. It is far above the morality of the conventional type. It is superior morality, meant for gaining true and lasting peace and joy. The importance of this morality of high order is not clearly understood by many students of Yoga. Yogic morality appears to be harsh or forbidding. Many aspirants resort to compromises in practicing *Yama*s and *Niyama*s. For example, some take *brahmacharya* as compatible with moderate sexual indulgences. Such compromises may seem quite reasonable from the worldly standpoint, but the true purpose of Yoga is not to keep a hold on this world, but to cross the ocean of worldly life. It is not that it is not possible to practice Yoga without giving up worldly enjoyments, but the progress of the practitioner is bound to stop at one stage or another, if he tries to make these compromises.

Another important point is that the virtues of *Yama* and *Niyama* have a **much wider scope and deeper significance** then their ordinary meaning. Each virtue is more comprehensive and deeper in its meaning than what it is generally considered to be. For example *ahimsa* means not mere non-violence, but it stands for the highest degree of harmlessness, not inflicting any injury, suffering or pain on any living being by thought, word or deed. To what degree of perfection the virtues of *Yama* and *Niyama* can be cultivated is shown in the eleven Sutras: II.35-45 of *Yoga Sutras* (See **last paragraph heading** in this chapter, below).

Thus the code of morality envisaged in *Yama* and *Niyama* is designed to serve as a very strong foundation for the higher Yogic life. The **main object of this supreme ethical code is to eliminate completely all mental and emotional disturbances** that characterize the life of an ordinary human being. Hatred, jealousy, dishonesty, deception, greediness, sensuality and possessiveness are some of the common vices of the human beings. As long as a human being is subject to these vices, he will not be free from mental and emotional disturbances. As long as these disturbances continue to affect the mind, it is not possible to undertake advanced practice of Yoga.

Let us now discuss briefly the significance of the five self-restraints prescribed under *Yama*.

Ahimsa

Ahimsa means **non-hurting** any living creature **in thoughts, words and deeds**. On the positive side, *ahimsa* leads to a spontaneous **love** toward all creatures. The Atman in all beings is one and the same. The cultivation of the attitude of *ahimsa* should start with the cultivation of positive thoughts and tender compassion toward all beings. Feelings of love and compassion purify the mind.

Sathya

Sathya means **truthfulness** in thought, word and deed. One lie inevitably leads to more lies. Lying is one of the easiest means of getting out of an undesirable situation. In avoiding one difficulty by lying, one creates many other serious difficult situations. Any effort to cover up falsehoods and deceptions creates guilty feelings and inner conflicts in the mind. Therefore truthfulness has to be practiced for keeping the mind free from such disturbances. On the positive side, this practice is absolutely necessary for developing the faculty of intuition and development of consciousness. Only a pure mind can reflect the ultimate Reality whose essential nature is love and truth.

Asteya

Asteya means not only **non-stealing** but also **refraining from misappropriation**, accepting bribes, taking credit for things one has not done or privileges that do not properly belong to him. The desire for what belongs to another is based on underlying feelings of inadequacy and jealousy. A student of Yoga should eliminate such attitudes by cultivating *asteya* and a sense of contentment and completeness.

Brahmacharya

Brahmacharya literally means, "to walk in Brahman," that is, being in awareness of Brahman alone. Such a state is possible, only if the mind is free from all sensual desires. Of all sensual desires, the sexual urge is the most powerful and the most destructive one. *Brahmacharya* is therefore translated as **celibacy** or complete abstinence from sex. Mere refraining from the sexual act is not continence. "Thinking of, talking about, joking, looking intently, secret talk, resolve, attempt and execution are the eight forms of sexual indulgence," say the sages.

Many Western writers have a liberal view. They take it to mean moderate indulgence within lawful wedlock. However, the real Yogic-way of life cannot be combined with self-indulgence, for sexual indulgence leads to the dissipation of vital energy that needs to be conserved for developing mental faculties essential for spiritual development. A married person, of course, may not be able to give up sex life all at once, but he has to give it up completely when he starts the practice of higher Yoga. Is anyone who is a slave of his passions really fit to embark on this Divine adventure?

Brahmacharya should not, however, be interpreted as suppression of sexual urges, for repression leads only to frustration and emotional upset. It means control of, and freedom from, all sexual cravings. The peace and bliss that accompanies Self-realization are far greater than any transient sexual pleasure. A Yogi whose goal is Self-realization would therefore overcome the obstacles of sexual cravings without any kind of suppression.

It must be clearly understood that the practice of higher Yoga requires complete abstinence from sexual indulgence, not only physical indulgence but even thoughts and emotions connected with sexual pleasures. No compromise on this point is possible.

The human energy that is expressed through sexual action and sexual thought, when checked and controlled, is easily transformed into *ojas*, a subtle spiritual energy. An unchaste person loses mental vigor and moral stamina. No wonder all the religious orders that have produced spiritual giants insist upon absolute chastity. There must be perfect chastity in thought, word and deed. Without it the practice of Raja Yoga is futile.

Besides, *Brahmacharya,* in its wider sense, stands not only for abstinence from sexual indulgence but freedom from craving for all kinds of sensual enjoyments, such as disturbing movies, novels, non-sattvic food, and mixing-up with persons of opposite sex.

Aparigraha

Aparigraha means "**non-possessiveness**." The tendency to accumulate wealth and worldly goods is very strong in human beings. Of course, as long as we live in the physical world we have to have a few things that are essential for the maintenance of the body. But we are not satisfied with the necessaries of life. We desire to have comforts and luxuries. We do not, however, stop even with all possible comforts and luxuries. We continue to amass wealth and possessions. There is no limit to our desires for material wealth. A story told by Sri Ramakrishna illustrates this:

The Jar of Desire can never be filled up

A barber, who was passing under a haunted tree, heard a voice say, "Will you accept seven jars full of gold?" The barber looked around, but could see no one. The offer of seven jars of gold, however, roused his strong desire for wealth, and he cried aloud, "Yes, I shall accept seven jars." At once came the reply, "Go home, I have carried the jars to your home." The barber ran home in hot haste to verify the truth of this strange announcement. When he entered the house, he saw the jars before him. He opened them and found all full of gold, except the last one, which was only half-full. He was not satisfied with what he got. A strong desire arose in his mind to fill the seventh jar also; for without it his happiness was incomplete. Therefore he converted all his ornaments into gold coins and put them into the jar; but the mysterious jar was, as before, unfilled. This exasperated the barber. Starving himself and his family, he saved some amount more and tried to fill the jar; but the jar remained as before.

So one day he humbly requested the king to increase his pay, as his income was not sufficient to maintain his family. The barber was a favorite of the king, and, as soon as the request was made the king doubled his pay. All this pay he saved and put into the jar, but the greedy jar showed no sign of filling. At last he began to live by begging from door to door, and his professional income and the income from begging – all went into the insatiable cavity of the mysterious jar. Months passed, and the condition of the miserable and miserly barber grew worse every day. Seeing his sad plight the king asked him one day, "Hallo! When your pay was half of what you now get, you were happy, cheerful and contented; but with double that pay, I see you morose, worried and dejected. What is the matter with you? Have you got 'the seven jars?" The barber was taken aback by this question and replied, "Your Majesty, who has informed you of this?" The king said, "Don't you know these are the signs of a person to whom the Yaksha consigns the seven jars. He offered me also the same jars, but I asked him whether this money might be spent or was merely to be hoarded. No sooner had I asked this question than the Yaksha ran away without any reply. Don't you know that no one can spend that money? It only brings with it the desire of hoarding. Go at once and return the money." The barber was brought to his senses by this advice, and he went to the haunted tree and said, "Take back your gold, O Yaksha." The Yaksha replied, "All right." When he returned home he found the seven jars had vanished mysteriously as they were brought in, and with it also had vanished his lifelong savings.

This story highlights how there is no end to the desire of a human being. Apart from the far-reaching imbalances that this human instinct causes in the social and economic fields, its effect on the life of the individual is terrible indeed. Consider the time and energy one spends in the accumulation of possessions and in maintaining and guarding the accumulated things. The worries and anxieties increase more than proportionately with increase in the accumulations. Then consider the constant fear of losing them. What a colossal waste of time and energy and mental agony all this involves! Is the precious human life meant for such wastage and misery? No one who is aware of the real purpose of life can afford to waste his limited resources of time and energy in this manner. So the Yogic aspirant does not waste his life in accumulating and maintaining possessions. He is satisfied with what comes to him in the natural course of life.

It is really what matters is not the quantity of our possessions, but our attitude toward them. A beggar, for instance, could be more attached to his begging bowl than a king to all his treasures. The danger lies not in having material possessions, but in becoming attached to them or in craving more.

No exceptions

A student of Yoga may sometimes be in doubt whether it is feasible or advisable to practice strictly the above five vows or exceptions can be made under special situations. Patanjali sets at rest all such doubts by making it clear that these great vows are universal, not limited by any consideration of class, place, time or circumstance. They can never be broken by any excuse (*Yoga Sutras*, II. 31).

Niyama

Meaning

The *Niyama*s are five observances to be followed. They are: *saucha* (purity), *santosha* (contentment), *tapas* (austerity), *swadhyaya* (self-study) and *Ishwara-praniddhana* (surrender to God) (*Yoga Sutras*, II.32).

How does *Niyamas* differ from *Yamas*?

Though both *Yamas* and *Niyamas* seem to have a common purpose, namely, the transmutation of the lower nature, there is a subtle difference between them. The *Yama*s are, in general, moral and prohibitive, while *Niyama*s are disciplinal and constructive. In the observances of the five vows of *Yama* the practitioner is required to react to the incidents and events in his life in a well-defined manner, but the number of occasions that will arise for putting the restraints to test will depend upon his circumstances. For example, if a Yogi lives alone in a jungle as an ascetic, there will hardly arise any occasion for testing any of the restraints. But *Niyamas* involve regular practices irrespective of the circumstances in which the practitioner is placed.

Saucha (Purity)

Saucha means "purity in both body and mind." What is purity? According to the Yoga Philosophy, the whole universe is a manifestation of the Divine Spirit. To the enlightened Saint who has attained Divinity, everything is pure and sacred. So the word purity in relation to our life is used in a relative sense.

It is easy to purify the physical body, but it is not so in the case of the mind. The purification of the body is achieved by refining it with pure and non-fat vegetarian diet, and avoiding meat, alcohol and smoking, which make the body unhealthy and utterly useless for the Yogic life.

Internal or mental purity is more important than external purity. The dirt of ego and *Tamasic* and *Rajasic gunas* makes the mind impure. The mind is purified by replacing all selfish and evil thoughts and emotions by pure and positive thoughts and emotions. The qualities that purify the mind, as given by the saint Ramanuja, are truthfulness, honesty, sincerity, compassion, harmlessness and non-covetousness. Cultivate discrimination and dispassion. Free the mind from lust, greed, anger, and attachment and aversion to the objects of senses. Wash out all the impurities from the mind and heart by prayer and *Japa*. Lord loves a pure heart. Kabir (14th century), a mystic of Banares of Northern India, says,

`Rama, the Lord, has possessed me, Hari, the beloved Lord has enchanted me,
All my doubts have flown like birds migrating in winter.
When I was mad with pride, the beloved Lord did not speak to me.
But when I became as humble as ashes, the Master opened my inner eye.
Dyeing every pore of my being in the color of love,
Drinking nectar from the cup of my emptied heart,
I slept in His abode in Divine ecstasy.
The devotee merges with the Lord like gold merging with its luster.
My Lord loves a pure heart.

The use of *mantra* and prayer is also a great purifier of the mind. They wash out all the impurities from the mind. Sincerity and perseverance are essential in cultivating mental and physical purity.

Santosha (Contentment)

Santosha or contentment is necessary for the Yogic aspirant for keeping his mind in a state of equilibrium. This state of mind does not depend upon one's material status. A beggar can be as content as a monarch. Our desires are insatiable and there is no end to desires. One desire leads to so many other desires. The mind is, therefore, in a constant state of agitation. It is constantly subject to anxieties, worries and disappointments. Besides, we are subject to all kinds of situations and encounters, and we react to them according to our nature. These reactions involve in most cases disturbances of the mind. The disturbed state of mind is not conducive to Concentration. So the Yogic aspirant has to transform this state of disturbance into a state of equilibrium and tranquility by a deliberate exercise of the willpower, Meditation and a conscious effort to accept whatever happens to him. He should develop the ability to withstand daily problems without being deeply affected by them, and to remain balanced in both pleasure and pain. This state of equanimity is the surest means for maintaining perfect mental peace.

Tapas (Asceticism)

Tapas means "to burn out" mental impurities. How these are burnt out? By accepting all the pain that comes to us. Such acceptance makes the mind serene, steady and strong. Thiruvalluvar in his sacred *Thirukkural* defines penance as bearing pain and suffering inflicted by others, and non-hurting other beings:

> *Urranoy nonral uyirkkurukan ceyyamai*
> *Array thavatthir kuru.*
> Bearing pain and not hurting other beings
> Are the hallmarks of penance (kural 261).
>
> *Sutachsutarum ponpol olivitum thunpanj*
> *Sutachsuta norkir pavarkku.*
> As fire goes on refining gold and makes it glow, so does
> The suffering purifies the doer of penance (kural 267).

Tapas also implies self-discipline and control of senses.

A simple life free from sensual indulgence, regulated fasting, observance of silence for a day every week, chanting the name of the Lord, and selfless service to fellow men and other beings—all these constitute *tapas*. *Tapas* helps to discipline the mind and to develop the willpower.

Lord Krishna in the *Bhagavad Gita* classifies penance or austerity into **three kinds**: bodily penance, penance of speech and mental penance. Using the body for the worship of God, and in selfless service is **bodily penance** (17.14). Speech that causes no excitement and is truthful, pleasant, beneficial and non-hurting is **penance of speech** (17.15). Serenity of mind, good-heartedness, purity of nature and self-control constitute **mental penance** (17.16).

In everyone's life there are hundreds of opportunities for *tapas*. As Swami Satchidananda writes in his commentary, even a cloth must undergo *tapas* to become clean. What will the laundryman do with our cloth? Will he fold it, put some sandalwood paste and a flower on it and give it back to us? No. First, he will soak it in boiling water with soap. Then he will beat it every way, and he will wash it and tumble and roll and squeeze it or wash it in a washing machine. After that he will dry it in a dryer or sun and iron it. Only then the cloth loses all its dirt and grime. It undergoes *tapas* to become pure. The laundryman has no hatred for the cloth when he does all these things to it. He only wants to make it pure. It is out of love that he inflicts pain.

The mind too must be tossed, washed, squeezed, dried and ironed to make it pure. If someone causes pain to us, we should not cause him pain in return, but rather thank him for helping us to purify ourselves. If we can think like this, we are real Yogis. Bearing another's insult or hurt with a serene mind is higher than any other austerity. The power to control the body and mind comes by *tapas*. If we accept everything with serenity, what can affect us? The best illustration for this is given in the "Song of the Mendicant" in the *Bhagavatam*.

The mendicant was badly hurt and insulted by some ignorant people. And he walked on, saying to himself, "Even if thou dost think another person is causing thee happiness or misery, thou art really neither happy nor wretched, for thou art the Atman, the changeless Spirit. Thy sense of happiness or misery is due to a false identification of thy Self with the body, which alone is subject to change. Thy Self is the real Self in all. With whom should thou be angry for causing pain if accidentally thou dost bite thy tongue with thy teeth?" This should be the attitude of a person who practices *tapas*.

In the name of *tapas*, some people practice all sorts of self-torture such as lying on a bed of nails. Such torture is not *tapas*. But self-discipline is absolutely necessary for spiritual progress. It will not obstruct one's normal life. Rather it will make one's life meaningful and peaceful (*See also* 3.1 Kriya Yoga, above)

Svadhyaya (Self-study)

This means **study of scriptures** that will elevate our mind and remind us of our true Self. They should be studied and understood deeply. We should contemplate, understand, assimilate and absorb the deeper and deeper thoughts. What is learnt must be put into practice. A deeper knowledge of the scriptures elevates our mind and expands our consciousness. But a mere theoretical knowledge is not enough (For further details refer 3.1 Kriya-Yoga, above).

***Ishwara-Pranidhana* (Surrender to God)**

The surrender to the Supreme Being is possible only with **infinite faith and dedication**. We should dedicate everything, our study, our practices, our thoughts, and our actions, and our heart and soul to the Lord. As Swami Satchidananda suggests, "We should pray 'I am Thine, All is Thine, Thy will be done.' 'Mine' binds; 'Thine' liberates.' .If we have 'mines' all over, they will 'undermine' our life. But if we change all 'mines' into 'Thine,' we will always be free and safe." True Yoga means dedication, giving, loving, selfless service, which will bring us permanent peace and joy and free our mind from its modifications.

Complete surrender to the Lord is one of the means for removing the veil of egoism. It transforms our "I" feeling and establishes communion with God. Of course, this is a gradual process and we have to consciously and sincerely practice self-surrender and devotion for a long time to attain success in it.

One who wants to retain his separateness as an individual cannot make a real self-surrender; his so-called surrender with that reservation does not destroy his ego-sense.

Once a young man came to Ramana Maharishi,[8] in a disordered state of mind. He believed he had had a vision of God, in which he was promised great things if he surrendered himself. He said he had done so, but that God failed to carry out His promise. He demanded the Maharishi: "Show me God and I shall chop off His head, or let Him chop off my head." Maharishi asked someone to read from a Tamil commentary on his own writings, and then made this remark: "If the surrender is real, then who is there remaining and able to question God's doings?" The young man's eyes were opened; he acknowledged that it was his own mistake and went away pacified. Self-surrender must be without any reservations and without conditions; there is no room in it for bargaining.

You achieve perfect surrender only when you give everything –

body, mind, senses and soul – completely to God without holding back anything. Your surrender should not be for any selfish motive, nor should you make a show of surrender. It should be genuine, real surrender with all your heart, with all your being. The Lord takes the responsibility for the redemption of one who takes total refuge in Him.

Perverse thoughts

Perverse thoughts like lust, greed, avarice, discontent, etc. are **opposed to *Yamas* and *Niyamas*.** Actions arising out of perverse thoughts such as stealing, causing injury, etc. may be done by oneself or caused to be done by another or approved when done by another; they are done either through greed, anger or delusion. Whether those actions are mild, moderate or intense, they result in endless pain and ignorance. So one should not let the mind to indulge in perverse thoughts (*Y.S.*, II.34). To tell a lie or cause another to tell one or to approve of another's lying – it is all equally sinful. Even a mild lie is still a lie. Every such vicious thought is stored up and will one day come back with tremendous power in the form of some misery. Perverse thoughts are dangerous and drive one to evil ways.

How can perverse thoughts be prevented or overcome? When perverse thoughts inhibit the *Yamas* and *Niyamas*, Patanjali suggests that **opposite thoughts** should be cultivated (*Y.S.*, II.33). For example, if the thought of hatred is in the mind, we can try to bring in the thought of love. In his commentary on this *sutra*, Swami Satchidananda says, "if we cannot do that, we can at least go to the people we love and in their presence forget the hatred. The easiest way is to change the environment. We can create a positive atmosphere by looking at a holy picture, or by reading an inspiring book or by meeting with a holy person or simply by leaving the disturbing environment. Another way to control negative thoughts even before the thought overpowers us is to think of its after-effects."

When perverse thoughts that are to be avoided become unproductive due to the cultivation of opposite thoughts, then the power acquired by it indicates the perfection of the Yogi.

Conclusion

The task of applying the *Yama*s and *Niyama*s to our daily life is difficult indeed, for they are strict codes of conduct. But the students of Yoga need not feel pessimistic and skeptic. They must have faith in themselves. Nothing is impossible if we have faith. Where there is a will, there is a way. If we have the determination, we can accomplish even extraordinary tasks. The initial step may seem difficult, but once we start practicing, we will gain confidence. We may fail now and then. Every failure is a stepping stone to success. We are not expected to attain perfection right away. We should regard the *Yama*s and *Niyama*s as ideals toward which we work with sincerity. We should make sincere efforts to observe them to the fullest extent possible. Even a small degree of success will reduce the intensity of emotional imbalances and mental disturbances. We can achieve greater and greater success by persistent practice and self-evaluation. Finally we can attain perfection and become fit for practicing higher Yoga.

Accomplishments of *Yama-Niyama*

What are the accomplishments of practicing the ten virtues of *Yama-Niyama*? When the practitioner reaches the state of perfection in practicing *Yama-Niyama*, he will experience the most fascinating manifestations. Patanjali points out these developments in *Sutras* II.35-45 of *Yoga Sutras*.

A Yogi who has cultivated **non-hurting** (*Ahimsa)* carries about him an invisible aura surcharged with love and compassion. In his presence all hostilities cease (*Y.S.*, II.35). He is inwardly attuned to all living creatures and automatically inspires confidence and love in them. Lord Buddha cultivated this virtue. Wherever he went he brought peace, harmony and friendliness. St. Francis is another great example for this.

A Yogi who has acquired **perfection in truthfulness** (*sathya)* becomes a mirror of Truth and whatever he says will come true (*Y.S.*, II.36).

To one firmly established in **non-stealing** (*Asteya)*, all wealth comes to him. If you are not avaricious and do not run after wealth, before long it runs after you (*Y.S.*, II.37).

To one firmly established in **celibacy** (*brahmacharya)*, energy is conserved and it is transformed into subtle energy for bringing about mental, moral and spiritual regeneration (*Y.S.*, II.38).

When perfection in **non-covetousness** (*aparigraha*) is attained, the Yogi acquires the capacity to have knowledge of his previous births. He directly sees the cause and effect relationship (*Y.S.*, II.39).

From the **physical purity** (*saucha*), disgust for one's own body and disinclination to have physical contact with others are developed (Y.S., II.40). This disgust is not aversion; it is only a natural disinterest in the body and relationship with another body.

From **mental purity** (*saucha)* arise pure *sattva*, cheerfulness, one-pointedness, control of senses and fitness for the vision of Self (*Y.S.*, II.41).

From **contentment** (*santosha*) arises supreme happiness, the Divine bliss. When there is no expectation, anxiety and desire, the mind is in its natural state of peace and joy (*Y.S.*, II.42).

When impurities are destroyed by **austerities** (*Tapas)*, sense organs and body become perfect, and the Yogi can use them for Yoga without any hindrance from them. He also gains occult powers (*Y.S.*, II.43).

By **study of spiritual books** (*svadhyaya*) communion with the personal God is attained. By constant study and consequent contemplation and enlightenment, we get a vision of our chosen deity ((*Y.S.*, II.44).

Surrender to God (*Ishvara-pranidhana*) can lead ultimately to Samadhi (*Y.S.*, II.45). This implies that surrender to God is practically an **alternate and independent path** of achieving the goal of Raja Yoga, namely Enlightenment or Self-realization. Surrender to God develops *Para vairagya* (supreme dispassion), breaks the bonds of the heart, removes egoism and personal desires and thus naturally and inevitably reduces the mind to a state of non-modifications, which is nothing but Samadhi.

Questions

1. Why have *Yama* and *Niyama* been included as the first stages of Raja Yoga?

2. What is the place of morality in Yogic life?
3. What are the significances of non-hurting and truthfulness?
4. How is celibacy important for Yogi?
5. Distinguish between *Yama* and *Niyama.*
6. Examine the significance of purity and contentment.
7. Is individuality compatible with self-surrender to God?
8. How do perverted thoughts affect the practice of *Yama* and *Niyama*?
9. How can the perverted thoughts be overcome or prevented?
10. What are the accomplishments of *Yama-Niyama*?

3.5: *Asanas*

Introduction

Patanjali has not described *Asanas* in detail. They were developed later by the exponents of Hatha Yoga. In the **manuals on Hatha Yoga**, such as *Hatha Yoga Pradipika, Gheranda Samhita* and *Siva Samhita, Asanas* are described in detail.

Asana means posture. Patanjali has defined posture as **holding the body in a steady and comfortable position** (*Y.S.*, II.46). Most of the *Asanas* tend to bring about very marked changes in the body and if practiced correctly and for a sufficiently long time promote health. A student of Yoga has to practice and perfect *Asanas*, for an unhealthy body is an obstacle for scaling the higher rungs of the Yoga ladder. Hatha Yoga is therefore necessary to ensure physical health and harmony that are prerequisites for Concentration and Meditation. It is an essential part of Raja Yoga.

Hatha Yoga

Hatha Yoga aims at mastery over the body with a view to securing mastery over the mind. The word "*hatha*" is derived from the roots *ha* (sun), and *tha* (moon). It means equalization and stabilization of the sun breath (the breath that flows through the right nostril) and the moon breath (the breath that flows through the left nostril). Through the regulation of the physiological process, Hatha Yoga releases the dormant energies of human personality.

The principal steps of Hatha Yoga are *Asanas* and *Pranayama*. *Asanas* consist of certain bodily postures such as sun worship, head-stand, shoulder-stand, etc. They are designed to stimulate the glands, vitalize the body and purify and strengthen the nervous system. *Pranayama* means regulation of the vital forces through breathing exercises.

Kinds of *Asanas*

Asanas are of two kinds: *Asanas* for Meditation, and *Asanas* for physical wellness. The *Asanas* suitable for *Pranayama*, Concentration and Meditation are *padmasana, siddhasana* and *sukhasana* and a few others. They are described later.

Asanas for physical wellness are many. They control specific muscles and nerves in the body and have specific therapeutic effects. They belong to Hatha Yoga. It is required to consult a manual on Hatha Yoga and to get personal instruction from a competent teacher.

The Need for *Asanas*

In Raja Yoga the method adopted for development of consciousness is based essentially on the control of the mind and the stoppage of mental modifications. The technique of Raja Yoga is, therefore, directed toward the elimination of all sources of disturbances to the mind, external and internal. One of the important sources of disturbance to the mind is the physical body. There is close connection between the mind and the body. They always act and react on each other. A student of Yoga must eliminate completely the disturbances arising from the physical body before tackling the internal disturbances. This is achieved through the practice of *Asanas*. When the body is fixed and kept in a particular posture for a reasonable time, it ceases to be a source of disturbance to the mind. The Yogi has to choose any one of the *Asanas* suitable for the practice of *Pranayama* and Meditation such as *padmasana* or *siddhasana* and then remain in that posture until he can maintain it without making any movement. Sitting in any posture becomes uncomfortable after a few minutes. If however that *Asana* is correctly chosen and practiced in the right way, steady and persistent practice will gradually eliminate all these minor discomforts that cause constant distraction to the mind.

Essential Features of *Asanas*

Patanjali has condensed all the essential features of *Asanas* in three *sutras* (II. 46 - 48). The essential features of *Asanas* are:

1. Posture should be **steady and comfortable** (*Yoga Sutras*, II. 46). These two, steadiness and comfort, are essential requirements of *Asanas*. Steadiness means fixing the body in one position without any movement, but there should be no rigidity, because rigidity makes the body tense. What should be aimed at is an ideal combination of immovability and relaxation. It is only then that it is possible to forget the body altogether.

2. In order **to achieve steadiness, two approaches** have been recommended by Patanjali (*Yoga Sutras*, II. 47). The **first** is the gradual slackening of effort. The keeping of the physical body in an immovable position for a long period of time requires great effort of the will and focus of the mind on the body. But this state of the mind is the opposite of what is aimed at. The mind has to be freed from the awareness of the body, not tied down to it in the effort to keep it in a particular posture. So the practitioner is advised to relax the effort gradually and transfer the control of the body from the conscious mind to the sub-conscious mind.

The **second** approach for securing steadiness is Meditation on *Ananta* (endless), the great Serpent that, according to Hindu mythology upholds the earth. *Ananta* is the symbolic representation of the force that maintains the equilibrium of the earth and keeps it in its orbit around the sun. This force is similar to one that works in a gyroscope used for maintaining an object in a position of equilibrium. Meditation or deep contemplation over that force tends to bring down that force in our life.

3. The most important **result of attaining perfection** in the practice of *Asanas* is **freedom from the disturbing reaction** of the pairs of opposites such as heat and cold, humidity and dryness (*Yoga Sutras*, II. 48). They distract the mind. Therefore the practitioner has to acquire the capacity of rising above them.

There are also other important benefits of practicing *Asanas*. They are:

4. The body becomes perfectly healthy and resistant to fatigue and strain.

5. The practitioner acquires fitness for the practice of *Pranayama* with ease.

6. Willpower is developed. It is an outward manifestation of spiritual power.

Meditative Postures

Meditation requires a posture that is comfortable and steady, one in which the head, neck and trunk are straight and in line. This means that the weight of the body should be distributed around its axis. If the spinal column is not erect, the aspirant's body begins to tremble after a few minutes and thus disturbs his mind; the trunk begins to form a curve, restricting the flow of vital energy through the spinal cord. As a result the gland centers are not energized by the circulation of blood, and the restricted blood circulation upsets the respiratory system. So in order to avoid these problems, the ancient Yogis have formulated some postures for practicing *Pranayama* and Meditation. They are *Padmasana* (lotus posture), *Sukhasana* (the easy posture) and others. All these postures involve sitting in an upright position with the spine straight.

The Yogis say that there is a subtle energy (*kundalini*) at the lower end of the spinal column. When one is spiritually uplifted this energy rises up in a channel within the spinal column. By keeping the spinal column erect the flow of this energy is not obstructed. The spine may be kept straight by lying flat or standing up, but it creates certain disturbances like sleepiness or restlessness. Therefore the sitting posture is prescribed for Meditation.

In the beginning stage, you can first sit in *Sukhasana,* but you should slowly try to proceed to the classical Meditative *Asana* such as *Padmasana*.. Persons with very stiff legs or who are infirm from any debilitating disease can practice Meditation on a straight-backed chair or lying on a hard bed, if absolutely necessary.

When you sit in a posture, imagine yourself as firm as a rock. The steadier you are in your *Asana*, the better is your Concentration.

After a regular practice for several months you will surely be able to sit for two or three hours at a time.

You can sit for twenty or thirty minutes in the beginning. You should progressively prolong the duration of sitting by one minute or two minutes daily. The ability to sit in *Padmasana* depends not only on the flexibility of the body, but also on the state of the mind. If you believe in your own mind you will eventually be able to sit in *Padmasana*, the mind itself will help to prepare the body.

Do not on any account use undue force or strain to sit in an *Asana*. If you find severe pain in the legs after some time in a Meditative *Asana,* slowly unlock the legs and massage them for a few minutes and then resume the posture.

Padmasana **(Lotus Posture)**

This is the most popular posture for Meditation. Its technique is:

- Sit on the floor on a rug or folded blanket or Meditation pillow.
- Fold one leg and place its foot on the top of the opposite thigh. The sole of the foot must be upward and the heel should touch pelvic bone.
- Fold your other leg and place its foot on the top of the other thigh. The spine must be steady and upright. Many practitioners find it comfortable to place a low cushion under their buttocks before assuming the posture.

The *benefits* of *Padmasana* are:

- It allows the practitioner to hold his body steady for long periods of time. Steadiness of the body brings steadiness of the mind.
- It directs the proper flow of *prana* from the bottom of the spine to the top of the head.
- It helps to clear up many physical, nervous and emotional problems.

Sukhasana

This is the simple cross-legged posture. It is the ideal posture for beginners who have difficulty in sitting in any of the classical meditative postures. Once the practitioner can comfortably do any of the other postures, *Sukhasana* should be disregarded.

The *technique* of *Sukhasana* is:

- Sit with legs stretched in front of the body.
- Fold the right foot under the left thigh.
- Fold the left foot under the right thigh.
- Place the hands on the knees.
- Keep the head, neck and back straight.

. The ***Asanas* for physical well-being** are explained in detail in the manuals on **Hatha Yoga**. *Asanas* should be performed on a carpet or folded blanket in a clean, quiet, ventilated room or in the open air. They may be performed either in the morning or in the evening. Warm bath before the *Asanas* is helpful, as it promotes circulation and reduces stiffness of the joints. The bladder and bowels must be emptied before practicing *Asanas*. At least four hours should have elapsed from the last normal meal. It is beneficial to begin the practice with a chant or prayer. This will center the mind, and the vibrations will calm down the mind.

All postures should be treated as Meditation in movement. All movement should be slow and controlled. *Asanas* should be performed only to the extent possible; over-straining should be avoided. After each posture, one should lie on the back in the corpse posture, and relax the body for one or two minutes.

The *Asanas* tone up the nervous, endocrine, circulatory, digestive, urinary and respiratory systems that are significant to physical health. Patience, perseverance and regularity ensure success. (For further details, consult *The Principles and Practice of Yoga of Meditation* by the author of this text or any manual on Hatha Yoga)

The second component of Hatha Yoga is ***Pranayama***. It means regulation of the vital energy (*prana*) through breathing exercises. It aims at the mastery over the vital forces flowing in the body.

Through the control of the breath and mobilization of vital forces, it endeavors to secure the release and free flow of the basic psycho-physical energy called *Kundalini* latent in the human system. (For details, *See* 3.6 *Pranayama, below*)

One who acquires success in Hatha Yoga enjoys vibrant health, youthfulness and longevity. He attains spiritual liberation and supreme bliss through the practice of Raja Yoga. Hatha Yoga serves as a means for developing the body as a fit and strong instrument for higher spiritual practice.

Hatha Yoga *Asanas* differ from other systems of physical culture. While other systems aim at merely developing a muscular body, Hatha Yoga aims at promoting the physical wellness and the health of the internal organs as well. Thus, complete physical harmony is achieved. It prepares the physical system for *Pranayama* and Meditation.

The chief defect of Hatha Yoga lies in its over-emphasis upon the body. The need for the development of the consciousness is not sufficiently recognized. Therefore it is absolutely necessary to **integrate Hatha Yoga with the major kinds of Yoga**, namely, the Yoga of Devotion, the Yoga of Meditation (*Raja Yoga*), the Yoga of Action and the Yoga of Knowledge.

Questions

1. What are the two kinds of *Asanas*? Why should they be practiced?
2. Explain the essential features of *Asanas*.
3. Describe the two common Meditative *Asanas*,
4. How does Hatha Yoga differ from other systems of physical culture?

3.6: *Pranayama*

Introduction

Pranayama is the **fourth step** in the eight-step Raja Yoga. It is not, as commonly understood, regulation of breathing. The word *Pranayama* is composed of two words: *prana* and *ayama*. *Prana* means "vital energy," and *ayama* means "control" or "restraint." Therefore *Pranayama* means the control of vital energy. *Prana* is the subtlest aspect in human beings and the universe.

According to the Indian philosophy, the universe is composed of two entities: *Akasa* and *Prana*. *Akasa* is the all-penetrating existence. Everything that has form is evolved out of *akasa.* It is *akasa* that has become the air, the liquids, the solids, the sun, the earth, the moon, the stars, the comets, the human body, the animal body, the plants and everything that exists. *Akasa* is so subtle that it cannot be perceived. It can only be seen when it has taken a form.

Prana

By the power of *prana* the universe is evolved out of *akasa.* Just as *akasa* is the omnipresent material of this universe, so is *prana* the omnipresent manifesting power of this universe. At the end of a cycle all tangible objects resolve back into *akasa*, and all the forces in the universe resolve back into *prana*. In the next cycle, out of *prana* is evolved every energy – motion, gravitation and magnetism. From thought down to physical force, everything is evolved out of *prana*.

Thought is the finest and highest manifestation of the *prana*. Conscious thought is not the whole of thought. There is also what is called instinct or unconscious thought, the lowest plane of thought.

If a mosquito bites me, my hand will strike it instinctively. All reflex actions of the body belong to this plane of thought. There is also another plane of thought, conscious thinking, reasoning and judging. Reason is limited. It can go only to a certain length, beyond that it cannot reach. Yogis find that the mind can function on a still higher plane, the Superconsciousness. "When the mind has attained that state, which is called *Samadhi*, it goes beyond the limits of reason and comes face to face with facts which no instinct or reason can ever know. All manipulations of the subtle forces of the body, different manifestations of *prana*, give a push to the mind, help it to go up higher and become Superconscious, from where it acts." (Swami Vivekananda, *Raja Yoga,* p. 38).

Prana – the Vital Force

The vital force in everything is *prana*. It sustains life. Whatever moves or works or has life is but a manifestation of *prana*. It is the power that supports the body and all its moving life forces. The *nadis* are channels for *prana*. When *prana* is in motion and flows through the *nadis,* consciousness arises.

Prana enters the body through the food we eat and the air we breathe. But *prana* is neither the food nor the air we breathe. Food and air are vehicles.

It is the *prana* that moves the lungs. The motion of the lungs draws in air. So *Pranayama* is not breathing, but controlling that muscular power that moves the lungs. That muscular power, which is transmitted through the nerves to the muscles and from them to the lungs, making them expanding and contracting, is the *prana* that we have to control through the practice of *Pranayama*. When this *prana* has become controlled, then all other actions of *prana* in the body, even muscles and nerves of the body, will slowly come under control. This the Yogi does through *Pranayama*.

Prana causes the senses of perception and organs of action to perform their respective functions. It is the force that holds together the elements of the body. It ignites the internal fire that maintains warmth and metabolism.

Prana is lost from the body in exhalation, excessive exercise, elimination of waste, the emission of semen, the process of childbirth and in times of great emotion. If *prana* is not regulated, the system does not receive an adequate supply of oxygen. The absence of appetite may also indicate an imbalance in *prana*.

If *prana* recedes from any part of the body, whatever be the causes, that part loses its power of action. At death, the *prana* exits through the eyes, ears, nose, navel, rectum, urethra or fontanel, leaving an impression at the site of its exit. Its tendency is to leave the body from the site where the mind dwells instinctively or where the inner most feelings reside.

Breath

Breath is an external manifestation of the force of *prana*. Just as the control of flywheel of an engine controls all other mechanisms in it, so the control of the external breath leads to the control of the gross, subtle, physical and mental aspects of our life. If the motion of the lungs stops, all other manifestations of the force of *prana* in the body will immediately stop.

Pranayama

Pranayama is one of the most important practices of Yoga science. It is concerned with the control of *prana*, the vital force in the body. This control is exercised through the regulation of the motion of the lungs. Though *prana* is different from breath, still there is a close connection between them. This connection enables us to manipulate the currents of *prana* by manipulating breath. *Pranayama* involves regulation of the three processes of breath: exhalation (*rechaka*), inhalation (*puraka*), and retention (*kumbhaka*), and thus it establishes control over *prana*.

The retention stimulates nerves, equalizes inhalation and exhalation, reduces the weight of the body and decreases the speed of thoughts. The Yogi who is capable of regulating the length of the exhalation and inhalation enjoys health and acquires spiritual power.

Just as the impurities of a metal are burnt away by fire, so the impurities of the mind and senses are burnt away by the control of *prana*, and thus *Pranayama* purifies the mind; it uncovers wisdom that is latent within each person. It is said that through *Pranayama*, the power of levitation is acquired, diseases are cured, spiritual energy is awakened, mental peace and joy, and mental powers are obtained.

The practice of *Pranayama* can be taken up safely, only as a part of full Yogic discipline with practice of *Yama-Niyamas* and *Asanas* under the guidance of a competent teacher.

Aims of *Pranayama*

The aims of *Pranayama* are:

1. To control the flow of *prana* so that the disturbing forces arising from within are eliminated and the mind becomes focused and one-pointed.
2. To oxygenate the blood and to prevent dissipation of the energy resulting from deoxygenation.
3. To burn up as much waste material as one can. Therefore one should inhale more air and retain the breath.
4. To introduce pressure into the system, maintaining a proper balance between the outside and inside pressures. When this balance is not maintained, then the nerves, mind and muscles are affected. The body trembles and the mind fails to function normally.
5. To control the thoughts. In the final stage, the breath of life is made perfectly serene, followed by its passage through *Sushumna nadi* in the spine to the crown of the head, where the Knowledge of Self (Super-consciousness) is attained.

(For Further details, refer the Chapters on ***Pranayama*** in **The Principles and Practice of Yoga Of Meditation** by the author of this text)

PATANJALI'S THOUGHTS ON *PRANAYAMA*

Meaning

After mastering the posture (*Asana*) *Pranayama* is practiced for regulating the flow of inhalation and exhalation of breath (*Y.S.*, II.49). This *sutra* describes *Pranayama* as the regulation of incoming and outgoing air. During this practice, the mind has also to be focused on the object of Meditation. That is, the fixing of the mind on the object of Meditation and the regulation of the breath should be made as a single effort.

The various ways in which the regulation of the breath can be practiced is described in the next *sutra* (aphorism).

The **first kind** is the regulation of inhalation and exhalation of breath.

The **second *Pranayama*** consists of **three** operations: **External Operation** (*Bahya-vrtti*), **Internal Operation** (*Abhyantara-vrtti*), and **Stationary** (*Stambha-vrtti*). After exhalation, keeping the air outside and not drawing it inside immediately is an external operation. Similarly inhalation and suspension of breath is an internal operation. All these operations are to be **regulated by space, time and number**, and are to be **long or subtle**. The **third operation** is total **stoppage of breath** (*Y.S.*, II.50). The words *Rechaka* (exhalation), *Puraka* (inhalation) and *Kumbhaka* (Retention), which are in current usage, were not used in ancient times. They were coined later. Therefore Patanjali did not use these words.

The external operation, internal operation and suspension – these three operations are practiced according to space, time and number, and become long or subtle. What do these terms mean?

"**Space**" means the **distance** covered by the exhaled or inhaled air. Space has to be taken in two senses—external and internal. The space between the tip of the nose to the point up to which the flow of breath is extended is the **external space**. The **internal space** refers to the space inside one's body, up to the region of the heart, and also from the heart to the entire body from head to foot.

The practice of *Pranayama* with observation of the distance covered by the exhaled air from the tip of the nose is an operation regulated by **external space**. The exhalation is gradually weakened. The **internal space** is perceived by **feeling**. When the inhaled air enters the lungs, it should be felt in the region of the heart. This is Pranayama with the observation of internal space.

Taking the region of the heart as the center, it has to be felt that prana is spreading all over the body from the heart during the inhalation. This purifies the nerves and energizes the body.

"**Time**" means the **duration** of inhalation, retention and exhalation. The observation of time may be done by silent repetition of a mantra like OM. If the mind is fixed on the flow of sound, then the conception of the passage of time becomes distinct.

"**Number**" refers to the **number of counting of the duration** of inhalation, retention and exhalation. The counting may be done by the repetition of OM, OM, OM or in bunches like "OM-OM, OM-OM, OM-OM-OM." Thus in one bunch seven repetitions of OM are made. One may start with five counts of inhalation and ten counts of exhalation. After a reasonable practice, the count may be increased to 10 counts of inhalation, 15 counts of retention and 20 counts of exhalation. The mind should be focused on the breath. The flow should be slow, smooth and even. The exhalation should be very slow, smooth and controlled.

"**Time**" also refers to the **number of times the exercise is repeated** in one session, say ten or twenty times in one session, the **number of times *Pranayama* is practiced** in a day. It is desirable to practice it three or four times in a day.

"**Long**" means exhalation or suspension or retention of breath for a **long time**. "**Subtlety**" means **gentleness** or **thinning** of inhalation and exhalation, and **effortlessness** during the holding of breath. *Prana* is a very subtle energy. There should not be any haste or strain in practicing it.

Fourth *Pranayama*

When the external operation regulated by space, time and number is mastered, it can be transcended by skill acquired through practice. Internal operation also, similarly regulated, can be transcended through practice. After proficiency is attained through practice, both these operations become long and subtle. Now one can go beyond the external and internal operations. Then arises a form of **automatic suspension** or stoppage of movement of breath. This is the **fourth Pranayama** (*Y. S.*, II.51). Suppression of movement (retention) with one effort, without considering space, time and number, is the **third Pranayama**. This is performed all at once. But the fourth Pranayama is performed after practicing external and internal operations with the observation of space, time and number, and then going beyond them all.

The **unintentional retention** occurs automatically in deep meditation. When the mind comes to a standstill, the *prana* automatically does the same. When we are absorbed in something we are reading, the breath stops. In deep Samadhi, breath stops for several hours. The Yogi, then, does not die, because there is no wastage of energy. It is being preserved.

The Benefits of *Pranayama*

Now Patanjali talks about the subtle benefit of *Pranayama*. When *Pranayama* is regularly practiced with perfection, the **karma** that shuts out discriminative knowledge **dwindles away**, and the veil of **mental darkness** that covers the inner light of knowledge is **removed** (*Y.S.*, II.52). Thus it has been said: "There is no *Tapas* superior to *Pranayama*, which removes the impurities of the mind and makes the light of knowledge shine."

Our real nature is the Self. But out of ignorance we identify ourselves with the body or mind, and become subject to the bondage of karmas. *Pranayama* weakens the veil of ignorance and the resultant karmas and latencies derived Thus it makes the light of knowledge shine.

As a result of the practice of *Pranayama,* the **mind becomes purified** and thus **becomes fit for Concentration** (*Y.S.*, II.53). It has been states in the *sutra* I.34 that the mind becomes calm through the regulated exhalation or retention of breath Only when the mind is free from disturbance, the focusing of the mind on the object of Concentration is possible.

Questions

1. What is *prana*? Why is it considered to be a vital force? Why should *prana* be regulated?
2. How is *prana* regulated?

3. What is *Pranayama*? What are its aims?
4. Explain the first and second kinds of *Pranayama* suggested by Patanjali.
5. Describe the fourth kind of *Pranayama*.
6. What are the benefits of *Pranayama*?

3.7: Pratyahara

Introduction

Pratyahara means **withdrawal of mind from senses**. It is the **fifth step** of Raja Yoga. Why should mind be withdrawn from the senses? Lack of control over senses leads to a chain reaction that ultimately destroys a person himself. Lord Krishna beautifully presents this fact in the *Bhagavad Gita*:

> "By constantly dwelling upon sense-objects, one develops attachment to them, from attachment arises desire, from desire is born anger." (2.62)
> "From anger arises delusion, from delusion loss of memory, from loss of memory one loses reasoning, and from the loss of reasoning, one perishes." (2.63)

To gain any success in Meditation, the mind has to be withdrawn from the external world and turned inward and focused on the object of Concentration. Besides, the mind has to be purified by cultivating moral and ethical values. All religious and ethical scriptures all over the world teach us moral values: "Be good," "do good," "do not steal," etc. The *Upanishads* contain many ethical injunctions such as: "Speak the truth; follow righteousness," "Truth alone prevails, never falsehoods," "Do not covet the wealth of others." Telling will not help us. We must know how to control our mind. But the mind is a slave of physical objects; it is restless and wandering; it is a storehouse of memories of past experiences, a stream of thoughts, desires and emotions. So when we sit and start Meditation practice, immediately a stream of thoughts, fantasies, memories and emotions will start flowing; and we will not be able to concentrate the mind on the object of Concentration.

Restlessness of Mind

The problem with our mind is that it is continually receiving data about the outside world through the sense organs: the ears, the eyes, the nose, the tongue and the skin. Innumerable vibrations from all kinds of objects are constantly bombarding the sense organs. Those sense impressions reach the mind through the brain centers. When the sense organs, brain centers and the mind join together, we perceive the objects. The mind is attracted by some of them and desires arise in the mind. Besides, thoughts go on arising from the memories of past experiences stored in the mind. From one thought a series of associated thoughts arise. For example, when a thought of forthcoming vacation arises in the mind, we start thinking about the various choices of spending the vacation, their pros and cons, the arrangements for travel and stay to be made, etc. We are also concerned about the current problems and the future. Thus a constant stream of desires, thoughts, anxieties and imaginations goes on flowing.

Pratyahara (Withdrawal from senses)

Thus the **main source of thought-waves** that go on disturbing the peace of the mind is **mind's contact with the senses**. Turbulent senses, like wild horses, will hurl down even an advanced Yogi towards the objects of the senses, preventing him from reaching the goal of final Liberation. The *Bhagavad Gita* says,

> "The mind, which follows the wandering senses, carries away his discrimination as the wind (carries away) a boat on the waters." (2.67)

The mind, which moves along with the wandering senses, destroys the discrimination of an aspirant and carries him away from his spiritual path.

Can the **mind's contact with the senses**, which is the source of disturbances in the mind, **be curtailed**? Actually we do this often. When we are absorbed in an interesting book we automatically lose the awareness of the surroundings. We forget the ticking sound of the clock or voice of the people in another room. How does this happen? This happens because of mind's disconnection with sense organs. This gives us a clue to get the mind cut off from the awareness of our environment. That is, if the mind's association with sense organs is disconnected or if the mind is under control, then the mind will lose the awareness of our environment. This withdrawal of the mind from the sense organs is called ***Pratyahara***. When the **mind is so withdrawn**, due to lack of contact with their corresponding objects, says Patanjali, the **senses**, as it were, **follow the nature of the mind**, i.e., like the mind that stops its activities, the **senses** also **stop their functions** (*Yoga Sutras*, II.54). Just as bees follow the course of the queen bee and rest when the latter rests, so when the mind stops, the senses also stop their activities.

Is *Pratyahara* Possible?

Certainly it is possible. We see it practiced by some groups of people. The faith healers teach people to deny misery, pain and evil. Their philosophy is roundabout, but it is a part of Yoga upon which they have unknowingly stumbled. When they succeed in making a person cast off suffering by denying it, they really use a part of *Pratyahara*, for they make the mind of the person strong enough to ignore the senses. Similarly, the hypnotists, by their suggestion, excite in the patient a sort of morbid *Pratyahara* for the time being. Of course the hypnotist's suggestion can act only upon a week mind.

The control of the brain centers of a patient done by a faith healer or a hypnotist leads to the ultimate ruin of the patient. It is not really controlling the brain centers by the power of the patient's own will, but it is, as it were, stunning his mind for a time by sudden blows which another's will gives to it. It is similar to controlling a fierce horse by giving a heavy blow on its head. The patient, whose mind is so stunned often, loses his mind gradually, instead of gaining the power of perfect control over it. Every attempt at control that is not voluntary and not made with the individual's own will, not only is disastrous, but also defeats its very purpose. The goal of each soul is freedom from the slavery of matter and thought, mastery over external and internal nature. Instead of leading toward that goal, every will-current coming from another only rivets one more link to the already existing heavy chain of bondage.

Whosoever asks anyone to believe blindly or drags people behind him by the controlling power of his stronger will, causes an injury to humanity. Therefore everyone should use his own will to control the body and mind himself. He who has succeeded in attaching or detaching his mind to or from the brain centers at will has succeeded in *Pratyahara.* For, according to Swami Vivekananda, *Pratyahara* means, "Checking the outgoing powers of the mind, freeing it from the thralldom of the senses." When we can do this, then we shall have taken a long step toward freedom, before that we are mere machines.

But it is not easy to control the mind. It is hard indeed to control it! The mind is compared to a maddened monkey that is restless by its own nature. Someone made it to drink wine. It became still more restless. Then a scorpion stung it. The poor monkey found his condition worse than ever. To complete its misery a demon entered into him. Are there words to describe the uncontrollable restlessness of that monkey? The human mind is like that monkey! "Incessantly active by its own nature, it then becomes," says Swami Vivekananda, "drunk with the wine of desire, thus increasing its turbulence. After desire has taken possession, comes the sting of the scorpion of jealousy at the success of others; and last of all the demon of pride enters the mind, making it think itself all important." You can imagine how hard it is to control such a mind!

The Possibility

The mind is like a naughty child; it does the opposite of what you want it to do. So if you try to detach it from sense impressions, it clings more intensively to them. Instead, sit for some time every day, and allow the mind to run on. Let the monkey jump as much as it can; you simply wait and watch. Until you know what the mind is doing you do not control it. Give it the rein. Many undesirable thoughts may bubble up. You may even wonder how it was possible for you to harbor such hideous thoughts! But how long can the monkey mind jump? You will find that each day the mind's vagaries are becoming less and less violent, that each day it is becoming calmer. You will find that, day after day, thoughts will be fewer and fewer, until at last the mind is under perfect control. But you must patiently practice every day. The mind, finding itself, being watched in this way, realizes its stillness and becomes still. When the mind withdraws into itself, the sense organs, like faithful dogs, instinctively imitate their master. They withdraw themselves from the respective objects (*Yoga Sutras*, II.54).

The *Katha Upanishad* compares the battle with the senses to a journey by a chariot drawn by horses. It says,

> He whose chariot is driven by reason, and who controls the reins of his mind, reaches the end of the journey, the supreme abode of the All-pervading." (1.3.9)

If the senses are allowed to turn outward, they grasp the pictures of the outside world and transmit those messages to the mind, making it restless. Turned inward, the senses find stillness by taking the form of the mind itself. The *Katha Upanishad* says:

> "When the five senses and the mind are still, and the intellect does not stir, that is the highest state." (2.3.10)
> This steady control of the senses is Yoga. Then one becomes undistracted, for Yoga comes and goes." (2.3.11)

You should not delude yourself into thinking that you have gained mastery even after a few years of practice of *Pratyahara*. In the *Bhagavad Gita* Lord Krishna warns us:

> "The turbulent senses do violently carry away the mind of even a wise man, though he is striving to control them." (2.60)

Any minute, there can be a slip. You should always be alert and vigilant. For example, if you observe strict celibacy, you should stay away from persons of opposite sex. You need not dislike them, but you should not freely associate with them, nor even to hug and kiss them. Although your mouth will call them "sisters or brothers," your senses will take the action in a different way.

The principal method for the practice of *Pratyahara* is **indifference to external objects**, and turning the mind toward contemplation. For a Yogi who has attained some success in practicing *Yamas* and *Niyamas*, *Pratyahara* becomes easier to practice, because he gains some control over the mind.

The Outcome of the Self-control of the Mind

The successful practice of *Pratyahara* gives you **complete control over the senses**. You are no longer their slaves. You become their master (*Yoga Sutras*, II.55). When the mind is controlled, the senses are controlled, for the senses derive their power from the mind. When the sense organs are under control, every muscle and nerve will be under control, because the organs are the centers of all sensations and of all actions. Therefore there is no need to employ other means. In the *Bhagavad Gita* Sri Krishna says,

> "When, like the tortoise which withdraws on all sides its limbs, a Yogi withdraws his senses from the senses-objects, then his wisdom becomes steady." (2.58)

A Yogi who is endowed with this power of withdrawal can enter into *Samadhi* even in a crowded place or a battlefield. If the functions of the senses can be stopped along with the suspension of the activities of the mind, whenever desired, that is the best form of senses-control. That is true freedom, and real victory. If you are free from your own mind and senses, nothing can bind you. Even an imperial power cannot bind you. You are not afraid of anything.

It is interesting to note how the **first five stages of Yoga** eliminate one after another the different **sources of disturbance** to the mind, and prepare it for the final struggle with its own modifications.

First, by ***Yamas* and *Niyamas*** are eliminated the **emotional disturbances** caused by moral defects in one's nature.

Second, by ***Asanas*** the **physical disturbances** are eliminated.

Third, one has to deal with the disturbances caused by the **irregular or inadequate flow of *prana***, the vital force. These disturbances are removed by the practice of ***Pranayama***.

Last, through ***Pratyahara,*** is removed the major source of **disturbances coming through the sense organs**. Thus is accomplished the external Yoga, and the practitioner now becomes capable of treading the stages of internal Yoga.

Questions

1. What is *Pratyahara*? Why is it necessary?
2. Why is the mind restless and wandering?
3. Is *Pratyahara* possible? How?
4. Is the control of one's mind by a faith healer or a hypnotist desirable? Give reasons for your answer.
5. Is it easy to control the mind? What is the possibility?
6. Why should one be alert even after gaining mastery over senses?
7. What is the outcome of the self-control of senses?

3.8: Concentration

Meaning

Concentration (*Dharana*) is the **sixth step** in Raja Yoga and the first one in the three-fold internal combined Yogic process of *Samyama* – Concentration, Meditation and *Samadhi*. Concentration means bringing the scattered mind to a **point of focus**. It involves freeing the mind from habitual association with desires and feelings, through exercises that reveal its prejudiced and wandering ways, and then focusing it upon a single object or idea.

In everyday life we concentrate while reading a book, while writing, while doing a job, while inserting a thread through the eye of a needle, while driving a car through a busy street and so on. This concentration is called external, for it is something in the external world that holds our attention. This is a muscular exercise.

Yogic Concentration (*Dharana*), according to Patanjali, is **fixing the mind on** a **particular point** or object or idea (*Yoga Sutras*, III.1). It involves focusing and holding the mind's attention on an object or idea. **Attention** is therefore required for Concentration. Attention is of two kinds, **voluntary** and **involuntary**. Voluntary attention is that which is directed toward an object or idea by an effort. It requires willpower, determination and mental training. Involuntary attention is spontaneous, and does not demand any training or willpower. Concentration requires **voluntary attention**. Continued attention leads to Concentration.

Attention plays a very great part in Concentration. It is the basis of will. When it is properly guided and directed toward the internal world for purposes of introspection, it will analyze the mind and illumine the very many astounding facts for you.

The force with which anything strikes the mind is generally in proportion to the degree of attention bestowed upon it. Moreover, the great art of memory is attention. Inattentive people have bad memories.

Attention is **focusing of awareness**. It is a rare faculty. Celibacy (*Brahmacharya*) wonderfully develops this power. Attention can be cultivated and developed by regular practice. All the great persons of the world have risen up through this faculty. It is through the power of attention that mind carries out all its activities.

There was an artisan who used to make arrows. One day he was very busy at his work. He was so much absorbed in his work that he did not notice even a big party of the monarch with his retinue passing in front of his shop. Such must be the nature of your Concentration when you fix your mind on God.

Concentration is a great instrument not only in spiritual life and Meditation but it is also very much desirable in practical worldly life to achieve success. The story of Arjuna, in a contest at archery after completion of his studies, beautifully illustrates this. At the termination of the courses of archery, the *Guru* Kripacharya wanted to test the skill of his hundred and six pupils. The target was the eye of a wooden bird hung from a branch of a tall tree by a thread in the air. One by one, the Kauravas and Pandavas came to hit the mark. At last came Arjuna. He stood on the assigned place, bent his bow and put the arrow to it. The Acharya now asked him what he saw? Arjuna said, "The head of the bird." The Acharya asked, "What is the target?" The reply was, "The eye of the bird." "Look closer, what do you see now?" "Only the eye of the bird." "Shoot." The arrow hit the eye. In the concentration Arjuna saw nothing but the eye. Concentration has to be so exact and minute in archery. So it is to be in all affairs of the world, and more so in Meditation.

Mind's wandering

In the beginning stage, when you sit for Meditation and try to concentrate on the chosen object of Concentration, you will find that the mind drifts away within a few seconds. This is a common problem for everyone in the beginning stage. This is the very nature of the mind, for the mind is nothing but a flow of thoughts. It is generally compared to a drunken mad monkey. No wonder it is constantly restless, and thought after thought goes on arising in it.

Mind jumps from one object to another object, just as a bird jumps from one twig to another, from one tree to another.

All along from the childhood, the mind has been trained to look externally, and indulge in endless thoughts of worldly objects, desires, aversions, emotions, memories, beliefs, daydreams and problems of life. So when you sit quietly and start to concentrate, the mind habitually runs to its old grooves. What else can you expect? You need not feel discouraged and helpless. Make determined effort and do regular practice to tame the mind and gently bring it under control.

In the initial stage of practice, subtle extraordinary sense perceptions may occur. They can give you some confidence, and encourage perseverance (*Yoga Sutras*, I.35). If you concentrate on the tip of your nose, you will experience an extraordinary smell of perfume. If Concentration is fixed on the tip of the tongue, you will get a nice taste; and so on. Such experiences are of no value in themselves, but they will give you confidence, and serve to prove what can be done with the mind.

Objects for Concentration

The object to be chosen for Concentration can be either a sacred name, a mystic *mantra*, the cosmic syllable OM, OM Shanthi, Hari OM, etc. or a form. The form may be that of your personal God such as Sri Krishna, Siva, Vishnu, Jesus or Buddha or your *Guru*. If you concentrate on a physical form, after a time you can create its mental picture. If you do not want a particular human form, then you can choose the visual image of the sun, the moon, a star, flower, flame or any other form you like, because the Lord is present everywhere in every form.

According to Patanjali, there are **six main objects** on anyone of which one can concentrate:

1. Inner Light: The ancient Yogis believed that there was a center of spiritual consciousness, called "the **Lotus of the Heart**," situated between the abdomen and the thorax, which could be revealed in deep Meditation. They claimed that it had the form of a lotus with an inner light. It was said to be "beyond all sorrow and worry," since those who saw it were filled with an extraordinary sense of peace and joy. From the earliest times, the masters of Yoga emphasized the importance of meditating upon this glowing Lotus (*Yoga Sutras*, I.36). The *Kaivalya Upanishad* says:

> "Brahman is higher than heaven, shines in the cave of the heart . . ." (3).
> "In a solitary place, seated oneself in an easy posture, with a pure heart, controlling all the senses, bowing with devotion to the *Guru*, Meditate on the lotus of the heart, pure and devoid of passion, in the center of which is the pure, the sorrowless, the inconceivable, the unmanifested, the blissful (Brahman)"(5,6).

In the *Chandogya Upanishad* we read:

> "Within the city of Brahman, in our own body, there is the heart and within the heart there is a small shrine in the shape of a lotus. Within it dwells that which is to be sought after and realized." (8.1.1)

The *Mundaka Upanishad* says:

> "Within the Lotus of the Heart, He dwells, where the nerves meet like the spokes of a wheel. Meditate upon Him as OM. May you be successful in crossing over the ocean of darkness." (2.2.6)

Concentrate on this glowing Lotus of the Heart, because it localizes our image of the spiritual consciousness toward which we are struggling.

2. The Heart of a Great Soul: You can concentrate on a great soul's heart that is free from passion (*Yoga Sutras*, I.37). Let your mind dwell on some holy personality, such as Lord Krishna, Buddha, Christ or Ramakrishna or a saint. Concentrate upon His heart. Try to feel the saint's heart has become your heart. Both Hindus and Christians practice this form of Meditation.

3. Dream Experience: Sometimes you may have **dreams of Divine beings**. If you have such dreams, remember them and let your mind dwell on them (*Yoga Sutras*, I. 38). In Indian scriptures we find many instances of such holy dreams and even *mantra* initiations are received in dreams. If you have not had any such dream, then imagine the peace of deep sleep, and concentrate upon it.

4. OM: Concentration on OM is one of the highest forms of Meditation (*Yoga Sutras*, I.28). OM is the highest name of God. It stands for the supreme Truth. It is infinite consciousness and bliss. The inner meanings can be studied in the *Mandukya Upanishad* and other scriptures. All Meditation and mystical practices should begin and end with OM.

5. *Chakras*: Fix the mind upon any one of the higher centers of spiritual consciousness called *Chakra*s (*See* 4.16 Kundalini Yoga, in *The principles and Practice of Yoga of Meditation* by the author of this text). Scriptures speak of seven centers of consciousness: lotuses at the navel, the organ of reproduction, the organ of evacuation, heart, throat, forehead and the crown of the head. When the mind is fixed on any one of the first three lower centers, it is attached to worldliness. A Yogi should not concentrate on those centers. He can fix the mind **on heart or any other higher centers** (*Yoga Sutras*, III.1).

6. Divine Form: You can also concentrate on your **personal God**, trying to visualize the form either within a lotus of a *chakra* or outside your own body (*Yoga Sutras*, III. 1). By meditating on the God's form, you will begin to develop Divine qualities.

7. Anything That Elevates: One of the most attractive features of Patanjali's philosophy is its universality. There is no attempt to impose any particular cult upon the spiritual aspirant. Patanjali clearly says, "You can mediate on anything that will elevate you." (*Yoga Sutras*, I.39).

Purpose of Concentration

The experiences of Yogis since ages have shown that one can control one's mind through practice. The mind is like a lake disturbed by the rising waves of thoughts. The practice of Concentration helps to still the waves and when your mind is stilled, you can see your reflection in the water of the lake and know your own true nature. According to Yoga Science, we are not restricted to the three states of waking, dreaming and deep sleep. There is a fourth state called *turiya,* the state of Superconsciousness. To attain it, you must regularly try to concentrate and bring your mind to a focus, after which through Meditation you can expand it to the Superconscious state. The purpose of Concentration is to gather together the dissipated energies of your mind, and to lead your concentrated mind to the state of Superconsciousness.

The effort of gathering the mental rays on a point is similar to gathering the rays of the sun through a lens on a sheet of paper. When the rays are sharply focused to a point on the paper, the concentrated solar energy produces ignition and the sheet of paper burns. In the same way, the concentrated mind is powerful and is capable of piercing through the deep layers of the consciousness.

When the mind is controlled and becomes free from desires or free from sense-objects, the state of Meditation (*dhyana*) is achieved. Concentration is therefore the master key that opens the gates of Meditation, because prolonged Concentration results in Meditation. In fact, it is difficult to discern the dividing line between the two.

Without Concentration the energy of the mind is dissipated in vague thoughts, worries and fantasies. A disciplined person expresses himself more clearly through Concentration; a person of ordinary intellect, with highly developed Concentration, is more creative then the highly intellectual person of poor Concentration.

Patanjali has given elaborate treatment to the science of Concentration in *Yoga Sutras*, for he realized its utility in calming an agitated mind. With steady practice the nervous system and the mind are relaxed, and the mind then becomes calm, steady and one-pointed. You are thus led to the Superconscious state in which you experience the Divine bliss.

Methods of Concentration

Concentration may be broadly classified into (a) **external** and (b) **internal**, depending upon the object of Concentration. The method chosen should be based on one's temperament, but once the method has been chosen, the aspirant should practice it faithfully for at least three months, only then he will begin to see the results.

In the **external method, the eyes are open** and **focused on an external object** such as a point, flower, steady flame of a candle, a picture of personal God or the *Guru*. Or one could use a mirror and gaze at the midspace between the eyebrows in the reflection. The gaze should be steady but there should be no strain on the eyes. The nasal gaze and frontal gaze are also effective methods for developing Concentration. In the nasal gaze the eyes are gently focused on the tip of the nose, whereas in the frontal gaze the space between the eyebrows is the point of Concentration. There should be no violent or forceful effort involved in this practice. The duration of Concentration can be from half a minute to half an hour. In the beginning one may concentrate on an external object.

In the **internal method the eyes are closed**, and the mind is focused (1) on a word or sound, (2) or on breathing, (3) or on a mental image, or (4) on a psychic center or *Chakra*.

There are many other methods of Concentration mentioned in the *Upanishads* and in Yoga manuals, but they are not elaborated upon and are revealed only to initiates by competent *Gurus*.

Yogic methods for developing Concentration are scientific and exact, and all of them involve conscious method of Concentration. Then in the second stage of Concentration, involuntary movements of the mind are brought under conscious control.

Preparation for, and the Process of, Concentration

The preparation for, and the process of, practicing Concentration are described below:

1. Select a definite time each day for this purpose; early morning and evening hours are the best time for Yoga practice.
2. Select a suitable place for sitting for Meditation. If you select a room in your living place, it should be quiet, clean and airy. The light should not be very bright, and the temperature should be moderate.
3. Concentration should not be practiced immediately after a heavy meal, as it causes discomfort and drowsiness.
4. Do not try to concentrate when you are physically or mentally tired.
5. Concentration is easy when the mind and body are relaxed and when the nerves have been purified by *Pranayama*. It is therefore desirable to practice some Yoga *Asanas* and deep relaxation first. Deep breathing with regulation of the breath stills the mind.
6. Sit in a comfortable posture. The postures suggested for Meditation (in 3.5 *Asana,* above) are suitable for Concentration.
7. The mind should be undisturbed and free from worldly worries and emotional problems. Assume an attitude of detachment. Gently close the eyes, withdraw the mind from sense objects, and say to yourself, "I am not the body, senses, and mind. I am the eternal Atman. How can the emotions and impulses disturb me? I am completely detached." With these positive thoughts calm the mind.
8. Try to be a mere witness to your mental activity. Let the thoughts come and go, and do not get involved in them.
9. Assert clearly that for the next few moments that you are concerned only with bringing the mind under the control of the inner Self.
10. Turn the mind to the object chosen for Concentration. (This has been discussed in a previous paragraph in this Chapter). Focus the mind upon the object.
11. Use your will in a natural way, but do not try to force attention.
12. In the beginning, practice Concentration for about ten minutes. After gaining some experience gradually increase the duration of the time period.

13. When the attention drifts away from the object of Concentration, and the mind indulges in distracting thoughts, the flow of breath is interrupted. If the breath is perfectly smooth, Concentration can be held on the chosen object, and the awareness can expand. Therefore, it is desirable to focus your attention on the breath and normalize it, and at the same time, bring back the attention to the object of Concentration. Whenever there is a drift, bring back the attention again and again to the object of Concentration. As a result of such repeated practice, the duration of focus gradually becomes lengthened, the repetition of *mantra* creates a new groove, and the mind begins to spontaneously flow into the new groove, and thus Concentration is strengthened.

Questions

1. What is Concentration? How does Yogic Concentration differ from ordinary concentration?
2. What is the importance of attention to Concentration?
3. What are the choices suggested by Patanjali regarding the object of Concentration?
4. Discuss the purpose of Concentration.
5. Distinguish between external and internal Yogic Concentration.
6. How would you prepare for Concentration?
7. Discuss the process of Concentration.

3.9: Meditation

Introduction

Concentration culminates in Meditation (*Dhyana*). Meditation is the primary technique of Yoga. Meditation is an **uninterrupted focus** of the mind on the object of Concentration, leading one to the highest level of consciousness and unto the goal of Self-realization (*Y.S.*, III.2). The uninterrupted focus is like the pouring oil from one pot to another. Meditation is a means for controlling the thought-waves of the mind. It is an inward journey from the gross, to the subtle, to the subtlest aspect of one's being, the Spirit or Atman. It is the final rung on the ladder of Yoga to attain illumination or *Samadhi*. In the *Bhagavad Gita* Lord Krishna says,

> "With the mind not moving towards any other thing, made steadfast by the practice of habitual meditation, one goes to the transcendental Supreme Being." (8.8)

The Yogi who meditates constantly without allowing the mind to wander to sensual objects reaches the Supreme Being. Krishna also says,

> "When thy intellect ... shall stand immovable and steady on the Self, then thou shall attain Self-realization." (2.53)

When the Yogi's intellect becomes steady without distraction and doubt and firmly established in the Self, then he attains Self-realization.

In Meditation the conscious mind is stilled. Through the withdrawal of the mind from senses, and through Concentration, fixation of the mind is achieved, and then Concentration flows into Meditation. This uninterrupted flow of Meditation leads to *Samadhi*.

Meditation is an inward journey to reach your own Self. Its aim is Self-realization, a direct vision of Truth. It is a way of going from the known to the unknown. It helps you to discover the ultimate unifying principle of the universe.

Theology says that God exists and that we should believe in Him. Philosophy says that we should know the relationship between the individual, the universe, and the creator. Meditation gives a direct vision of God in the temple of the body. A meditating Yogi does not have to search, roam or wander in pursuit of God; he finds Him within; he realizes his Divine nature.

Types of Meditation

Meditation is of various types according to the object chosen for Concentration. It can be **concrete** or **abstract** or **symbolic**. Meditation on a physical form such as a candle flame or a deity is concrete. Meditation on an idea or sound is abstract. Meditation on a physical form, which symbolizes a spiritual idea is a symbolic Meditation, e.g., meditating upon a Divine form with the feeling that you are lifting yourself to the heights of infinity or meditating upon a sound symbol like OM with a profound understanding of its significance.

Japa

Japa is the repetition of a *mantra* or Divine Name during the practice of Meditation. The Sanskrit root "*pa*" stands for that which removes or destroys all impurities and obstructions, and the root "*ja*" stands for that which puts an end to the cycle of births and deaths. Therefore, *Japa* is an indirect means for Liberation. By destroying obstructions to knowledge, *Japa* paves the way for Liberation.

The repetition becomes a technique for keeping our mind focused for a length of time, and also for helping the mind to gain certain depth. Between two successive repetitions, there is an interval with no form or shape. This is what we call peace or silence.

In a Sanskrit text *Pancadasi*, the mind is likened to a dancer on a lighted stage. The dancer portrays a variety of aesthetic sentiments – love, helplessness, anger, cruelty, fright and so on. The light on the stage lights up the dancer in all of her moods and forms, and when she makes her exit, it lights up the empty stage. Whether the dancer performs various dance forms or leaves the stage, the light remains uninvolved. It merely illumines.

The light itself is not a "doer," much less an "enjoyer" of the dance. It also does not light up the stage as one of its jobs. The nature of light is to illumine and it illumines. Therefore, the light has no doership. Similarly, when we have a thought and it goes away, what remains is silence, which is likened to the empty stage without a dancer.

In *Japa*, being aware of the interval is as important as the repetition, because it is the interval that reveals our true nature—silence, awareness.

Once we are committed to mentally repeating a given *mantra*, our mind automatically goes to *Japa* whenever it is free. Just as water draining from the mountains creates new ravines, *Japa* creates a new track towards which the mind repeatedly goes. In this way, *Japa* becomes a way of keeping the mind meaningfully occupied.

Effects of *Japa*: *Japa* enables us to replace distracting thoughts with the chosen *mantra*. It eliminates the chain of thinking and is helpful in gaining mastery over our mind. Lord Krishna, in the *Bhagavad Gita* says,

> "Among the various forms of rituals and many means through which I am invoked, I am *Japa*." (10.25)

This shows the greatness of *Japa* as a technique of Meditation. God-realization is possible, only when the mind is controlled and purified. Constant *Japa* arrests the restless nature of the mind. It conserves all the physical and mental energies for removing the veil from the indwelling Truth. The *mantra* or the Divine Name is an unfailing key that unlocks the gates of the heart, permitting an outflow of immortal love, wisdom and power:

> "God's all-powerful Name," says Swami Ramdas,[9] "takes the aspiring soul to the highest summit of Truth. What is required first is an absolute faith in the greatness and potency of the Name, which comes only by the Grace of the Lord. When the Name becomes the sole mainstay and refuge of the aspirant who thirsts for the highest goal of life—God-realization—he or she marches towards the goal not only in rapid strides but also with a heart filled with courage and cheerfulness."[10]

When we do *Japa,* we must be aware that we are repeating the Name of One who is within us. Without it, the repetition becomes merely mechanical and does not help us in any way. When we tune ourselves with the Name, we tune ourselves with God. We must keep this central fact in our mind when doing *Japa.* Then the practice will gradually make us aware of Divine existence within us. This idea grows into an experience when we actually feel His presence. As we become aware of Him within us, we become aware of Him also without us. We feel His presence everywhere. Of all the disciplines for controlling the restless mind, none is so easy and effective as *Japa.* But some are not able to repeat the Name continuously although they desire to do so. It is because of their greater love for the perishable objects of the world.

As we think so we become. So if our mind is fixed with an intense love for the Divine Name, this love will automatically enable us to repeat the Name constantly. When thus the mind is absorbed with the love of God and filled with the music of His Name, we will experience the Divine ecstasy. When we are in this Joy, Swami Ramdas says, the vision of God will flash out. Thus *Japa* is the easiest and best method by which you can purify yourself preparatory to God-experience. Swami Ramdas affirms that no other spiritual practice could so easily grant you purity—absolute freedom from lust, greed and wrath.

Japa is an all-inclusive and all-sufficient practice by which you can be ever in tune with God and finally merge yourself in Him. *Japa* is thinking of God constantly to the exclusion of everything else. God-thought keeps away every other thought. This thought also ultimately disappears, affirms Swami Ramdas, and you realize your identity with God.

Meditation in Monism (Vedanta)

Monism is the original form of the Vedanta, based on the intuitive experiences of the sages of the *Upanishads*. In Monism the ultimate Reality is One only and it is called Brahman. In Monism the practical procedure consists in denying the absolute existence of body, mind and ego, as well as the universe, which are only relatively real, and arriving at the One Absolute Reality, Brahman. The Self of all beings is identical with Brahman.

Monistic Meditation is the Meditation on the Absolute. As the Absolute is beyond form and beyond attributes, this type of Meditation is the most difficult of all. It is not thinking of the mere formlessness such as a vacant space, the blue sky or the shoreless ocean. It is only thinking of the Infinity of the Spirit, called *Chidakasa*, the Void beyond all forms and attributes. This is the negation of everything conceivable or expressible and it is described in the *Upanishadic* words, "not this," "not that." This is inconceivably higher than the infinity of the universe and the mind, because both the universe and the mind are limited; but the Infinity of the Spirit, the Atman or Brahman is the true Infinity, the true Absolute of Monism.

Meditation on OM

The *Upanishads*, the basic texts of the Vedanta Philosophy, assert the practice of Meditation on OM as a dynamic means for attaining Self-realization. The *Mundaka Upanishad* says:

> "Pranava (OM) is the bow. The individual soul is the arrow, Brahman or the Absolute is the target, O Aspirant, hit the target with persistent alertness, and just as the arrow becomes one

with the target, so you are united with Brahman (the Absolute)" (2.2.4).

Patanjali enjoins,

> "Pranava or OM expresses God (*Ishwara*); It must be repeated with Meditation upon its meaning. Hence comes the Knowledge of the Atman, and destruction of the obstacles to that Knowledge" (*Y.S.*, I..27, 28, 29).

Lord Krishna proclaims in the *Bhagavad Gita*:

> "I am Pranava (OM) in the *Vedas*." (7.8).
>
> While meditating upon OM (AUM), you should understand the implications of 'A' and meditate: "I am all this physical universe, as far as 'A' extends."
>
> Then you must negate the idea of the physicality of things, and enter into the plane of the mind and assert your Meditation on the 'U' aspect of OM: "I am the Cosmic Mind, and all minds are waves in the ocean of the Cosmic Mind."
>
> You must transcend the mind as well, and meditate upon the unmanifested *Maya* (the causal plane of the universe) by asserting the 'M' aspect of OM: "I am the Cosmic Self, the ruler and sustainer of *Maya*; I am the Spirit, the prompter of the operations of *Prakriti* (Nature), I am the source, sustenance, and the termination of the world-process."

Further you must go beyond the three planes, physical, astral and causal, and attain the revelation of the Self or Brahman (without attributes) denoted by the sound OM.

Meditation upon the Self or Atman

The ultimate goal of Meditation is to experience the Self or Atman. The Self is pure Consciousness, and the experience of Self is a state of absolute knowledge and bliss. It is variously called *Samadhi*, *Nirvana,* Cosmic Consciousness and so on. When the mind is completely centered for an extended period of time without any distractions, you become aware of the Atman.

The Self within us is enveloped by three bodies. The outermost is the physical body, composed of gross matter. Beneath this is the subtle body, composed of subtler counterparts of gross matter. The innermost body is the causal body that has been formed out of our actions as per the Law of Karma. In normal waking consciousness all the three bodies exist between you and Atman, but during the dream state the physical body is removed, and in the state of dreamless sleep only the causal body shrouds Atman. You are therefore closest to Atman in the state of dreamless sleep, but no memory of the experience remains when you wake up, for this body is difficult to penetrate. That is why you cannot attain realization of Atman merely through dreamless sleep. One of the methods of Meditation brings back into the waking state the experience of the state of dreamless sleep, and it involves dwelling upon, and trying to strengthen, the sense of peace that lingers into the waking state after experiencing the dreamless sleep.

In the *Upanishads*, it is stated that the Atman dwells within the lotus of the heart, but here heart does not refer to the heart of flesh, but to the heart center (*Anahata Chakra)*. In the *Katha Upanishad* the Atman is described as a smokeless, pure, illumined flame the size of the thumb in the shrine of the heart (2.1.13). In their deeper states of Meditation Yogis experience the lotus of the heart where, bathed in its inner light, they experience Divine bliss.

One of the most advanced methods of Meditation is Meditation upon the lotus of the heart after you have made your mind steady through Concentration. Then, withdrawing from the objects of the senses, you enter the shrine of Atman and, meditating on it, transcend the body and experience a higher Knowledge. Finally you attain *Samadhi*, in which you achieve Oneness with the Atman.

Meditation in Raja Yoga

According to the different levels of his mental development, a Yogi may meditate upon a gross object, or a sound (*mantra*) or a subtle mode of mind as he advances on the ascending rungs of Yoga.

Meditation leads to *Samadhi* wherein the Yogi rises to a mystic expansion, transcending the subject-object relationship involved in the process of Meditation. He then experiences his true nature, the eternal Self.

Meditation in Bhakti Yoga

A devotee focuses his mind on the form or name of the chosen personal God (Lord Rama, Lord Krishna, Shiva, Lord Buddha or Jesus Christ) invoking the Divine presence in his heart. It is done during *Japa* or repetition of Divine Name or *Mantra* of the Deity. It involves thought of God to the exclusion of all others (*Narada Bhakti Sutras,* 9). The mind is to allow only the thought of God, and exclude all others. All words and actions are also dedicated to the Lord. Great pain results when the mind forgets the Lord (*N.B.S.*, 19). Should the mind stray a bit, it is brought back to the thought of God. True Bhakti is unceasing remembrance of God (*N.B.S.*, 36). At all times with heart and soul, God alone is to be remembered (*N.B.S.*, 79). So Bhakti Yoga is exclusive meditation on God only. When it becomes more intensive the devotee gets the vision of God. (*See* the subheading *Japa* in this Chapter, *above*).

Meditation in Jnana Yoga

An aspirant listens to the exposition of Vedanta Philosophy by the *Guru*; he reflects upon the teachings until his intellect is able to grasp the implication.

When he understands the nature of the Self as the Reality beyond the body, mind and sense, ego and the sense of individuality, he commences the Meditation on "I am Brahman" (*Aham Brahma Asmi*). This is called *Nididhysana.* He withdraws the mind from the objects of the world and purifies his mind. He, then, begins to experience the Self in varying degrees of expansion, when he goes beyond the mind, his "I consciousness" dissolves in the experience of the Universal Reality.

Just as a Bhakta has to constantly remember God, a Jnani has to directly to dwell on the Atman itself, i.e., on Self-awareness. *Gita* says, "Keep the mind on the Self itself and on nothing else" (6.25).

Brahma Sutras, the ancient scripture on Vedanta, also states, "For one who sets his mind on Self, freedom (*Moksha*) is the result" (1.1.27). His mind has to join the Self" (1.1.19). "By the meditation on Brahman (pure Self), the hidden glory of the Self reveals itself" (2.2.5). "The *Niddhyasana* has to be continued (till the Self-experience dawns)" (4.1.1). But "Meditation shall not be on any symbol, for Self is not that" (4.1.4). Thus there are two stages in Jnana Yoga. First, an aspirant attains intellectual conviction on the existence of the eternal Self through reflection upon the teachings on Vedanta, discrimination of the Real from the unreal, and dispassion. Second, he constantly meditates on the Self till the experience of the Self is attained.

A Jnana Yogi may also meditate upon *Soham*, a Vedantic *mantra* for Meditation. It means "That am I."

He focuses the mind on center *(Ajna Chakra)* between the eyebrows, associating the sound "*So*" with the incoming breath and "*ham*" with outgoing breath. He should feel that the incoming breath creates the sound "*So*" automatically, and the outgoing breath creates "*ham*." Along with the mental visualization of the sound associated with breathing, he should try to assert the meaning of these two words.

A Jnana Yogi may also meditate on the Self by adopting a symbol of space in the region of his heart. Within the space he visualizes the abstract presence of the infinite Self. He may meditate upon the nature of the Self, reflecting upon the teachings, "I am not this body, I am not this mind, nor the senses . . . I am immortal Self."

Indispensability of Meditation for God-Realization

Atman is the storehouse of all powers. Mind is only an instrument of Atman. It should be properly disciplined. Just as you develop the physical body through physical exercises or *Asanas*, you will have to train the mind through Meditation. Then only the gross mind will become subtle. To keep the mind charged with the power of supreme wisdom, you must keep it always in touch with the light of Knowledge of Self through constant and intense Meditation. You must keep up an unceasing flow of the Cosmic Consciousness.

Leading a virtuous life is not by itself sufficient for God-realization. Concentration of mind is absolutely necessary. A good, virtuous life only prepares the mind as a fit instrument for Concentration and Meditation. It is Concentration and Meditation that eventually lead to Self-realization.

God is within you. He is seated in the Lotus of your heart. You will have to seek Him through Meditation with a pure mind. All the visible things are illusion (*Maya*). *Maya* causes havoc through the mind. Meditation is the only way to conquer *Maya*. Buddha had conquered *Maya* and mind through deep Meditation only. Solitude and Meditation are two important requisites for Self-Realization.

Questions

1. What is Meditation? What is its aim?
2. How is Meditation an inward journey?
3. How do you classify Meditation?
4. Describe the meaning and process of Japa.
5. Explain the Meditation in different kinds of Yoga.
6. What are the implications of Meditation on OM?
7. Describe the technique of Meditation upon the Self.

3.10: Samadhi

Meaning

Concentration culminates in Meditation, and Meditation in Samadhi. Samadhi is the final stage of Yoga. It means perfect Concentration or absorption. **When the state of Meditation is so deep that only the object (of Meditation) shines, without any trace of reflective thought, not even the awareness of oneself (who is meditating). That state is called Samadhi** (*Y.S.*, III.3). In the lower state of Meditation one is conscious of **three forms of feelings:** (1) the object meditated upon, (2) the process of Meditation, and (3) oneself that is meditating. But **when Meditation deepens, the mind gets lost in the object of Meditation.** How this state is reached?

As a result of observing the virtues of *Yama* and *Niyama* and regular intense practice of *Pranayama, Pratyahara* and concentrated Meditation, a Yogi's mind becomes so pure that he develops the ability to get absorbed in the object of Meditation. This ability is not confined to the object of Meditation only. **He can focus the mind on any object** – whether it is as subtle as an atom or as vast as the Cosmic Mind itself (*Y.S.*, I.40).

Just as a pure crystal assumes the color of any object placed near it, so the mind, when cleared of totally weakened modifications, identifies itself with the object of Meditation. **In this state one who meditates, the mind and the meditated—all become one.** The **mind** loses its awareness and **becomes identical with the object of Meditation** (*Y.S.*, I.41). How does this happen? How do the modifications become waned or weakened? The weakening of the modifications is caused by the cultivation of one thought –

the object of Meditation at the cost of all others. **When one alone is cultivated, all other impressions become weaker and weaker.** For example, if you concentrate on the development of the brain alone, you are apt to ignore the other parts of the body. Nowadays, the people at the shops and stores are not able to do even simple calculations orally, because they use calculators or computers for every calculation. Considering the present trend of using machines and gadgets for every work, instead of the limbs of the body, it may not be a wonder if the future generation have only a big head with little limbs like the roots of potato as described by H.G. Wells in a story.

This is not only true of the physical body, but also true of the mind. **If you constantly meditate upon one idea, all other thoughts and desires will gradually die out.** In the ancient Hindu scriptures there are stories to illustrate this point. For example, there is the **story of Valmiki**, the highway robber, quoted by Swami Satchidananda in his commentary on *Yoga Sutras:*

> Sage Narada was passing by. As usual, Valmiki accosted him and said, "Hey, what do you have in your pocket?"
>
> "Oh, I don't even have pockets, sir."
>
> "What a wretched man. I've never seen a man with nothing. You must give me something, otherwise I won't spare your life."
>
> Then Narada said, "All right, I will try to get something for you; but don't you think it's a sin to harm innocent people?"
>
> "Oh, you swamis talk a lot about sin. You have no other business, but I have to maintain my wife, children and my house. If I just sit and think of virtue, our tummies have to starve. I have to get money somehow, by hook or crook."
>
> "Well, all right, do it. If that's your policy, I don't mind. But you say you must feed your wife and children by hook or crook. You should know that it is a sin and you will have to face the effects of it."
>
> "Well I don't bother about that."
>
> "You may not bother, but since you are committing sins for your wife and children, you'd better ask them whether they are

willing to share the effects of the sins also."

"Undoubtedly they will. My wife always says we are one, and my children love me like anything; so, naturally, all I do for their sake will be shared by them."

"Well may be so, but don't just tell me. Go find out for sure."

"Will you run away?"

"No."

"Okay. You stay here. I will run there and find out."

So he ran to his house and said, "Hey, this man asked me a funny question just now. He says that I am committing sins, and certainly there's no doubt about that. But I am doing it for your sake. When you take a share of the food, will you take a share of the sin also?"

"It is your duty as a husband and father to maintain us. It is immaterial to us how you do it. We are not responsible. We didn't ask you to commit sins. You could do some proper work to bring us food. Anyway, that's your business and your duty. We are not going to bother whether it's right or wrong. We won't take a share of your sins."

"My God! My beloved children, how about you?"

"As Mommy says, Dad."

"What a dirty family. I thought you were going to share everything with me. You are going to share only the food and nothing else. I don't even want to see your face!"

He ran back and fell at Narada's feet. "Swamiji, you have opened my eyes. What am I to do now?"

"Well. You have committed a lot of sins. You have to purge them all."

"Please tell me some way."

So Narada gave Valmiki mantra initiation. "All right. Can you repeat, 'Rama, Rama'?"

"What is that? I've never heard of it; I'm just an illiterate person. I can't repeat it can you give me something easier?"

"Oh, what a pity. Let's see, look at this." He pointed at a tree. "What is it?"

"It's a *mara* (tree)."

"All right. Can you repeat it?"

"Sure. That's easy."

"Fine. Sit in a quiet place and just go on repeating, 'Mara mara mara…'"

"Is that all? Will that save me from all my sins?"

"Certainly."

"Well sir, I believe you. You have already enlightened me quiet a lot. You seem to be a good swami. I'll begin right here and now. I don't want to waste any time."

So he just sat under a tree and went on repeating, "Mara, mara, ma ra ma rama, Rama, Rama (God's Name)…" See? Mara mara soon became Rama Rama. He sat for years like that until at last an anthill was formed completely covering his body, because he was so deeply absorbed in repeating the mantra that he forgot everything else, even his body. This is what happens in Samadhi. After a long time somebody just happened to disturb the anthill, and the saint Valmiki emerged. Later, he got the divine vision of Lord Rama's life and wrote the entire epic story of the *Ramayana*. His *Ramayana* continues to influence the culture of Hindus since then.

What is the lesson that we learn from this story? When he concentrated on that mantra, he forgot everything else. All his sins slowly dried up for want of nourishment and died. If a plant is not watered, it slowly withers and dies. Our habits will also wither and die if we do not give them an opportunity to manifest. Similarly **the modifications in the Yogi's mind become powerless by the cultivation of one thought, mantra** (the object of concentration).

In the **perfect state of Samadhi** the **mind rests in the Self,** free from any thought wave, then it is said, "the Yogi is united" (i.e., he has become one with Brahman) (The *Bhagavad Gita,* 6.18). Just as a river, on merging into the sea, loses its name and form, and becomes one with the sea, so does the mind of the Yogi, who meditates, merges into the Self. Then there is no mind, no body-consciousness, only Self exists.

Kinds of Samadhi

Samadhi is broadly classified into (a) **Samprajnata (with form)**

Samadhi and (b) **Asamprajnata (without form) Samadhi.** Samprajnata Samadhi may be practiced on a gross object or on a subtle object, or on mind's own joy, or on the feeling of pure I-ness. Accordingly it is classified into (1) Savitarka Samadhi, (2) Savichara Samadhi, (3) Sananda Samadhi, (4) Sasmita Samadhi. As against Savitarka Samadhi, there is (5) Nirvitarka Samadhi. Similarly as against Savichara Samadhi and there is (6) Nirvichara Samadhi. All these are lower forms of Samadhi with seed or form. Asamprajnata Samadhi is higher form of Samadhi without seed or form.

Samprajnata Samadhi: This is a Samadhi accompanied by **reasoning** or deliberation (*vitarka*) or by **reflection** (*vichara*) or by **bliss** (*ananda*) or by a **sense of I** (*asmita*) (Y.S., I. 17). The **object of Meditation** (on which the mind is concentrated) may be a **gross object** (like a flower or the symbol of one's chosen Deity) or **subtle object** (like red, blue, love) or **blissful feeling** or **pure I-sense**. Whatever may be the object on which the mind is concentrated, in this Samadhi, **duality** – the person who meditates and the object of Meditation – prevails. The mind does not completely merge in the Self.

In all forms of Samprajnata Samadhi the mind still contains the seeds of desires in the form of subtle subconscious impressions. They can come into the conscious mind when proper opportunities arise, and pull the Yogi into the worldly experience.

Savitarka Samadhi: This is the **first stage** of the lower forms of Samadhi. This is reached by meditating upon a **gross object** like a rose with awareness of its name, form and knowledge (Y.S., I. 42). In this stage the Yogi is able to develop a deep Meditation on the chosen gross object, but he is not yet able to go beyond the mental associations caused by the name, form and knowledge of the object. When he succeeds in separating the object from the mental associations, he enters into Nirvitarka Samadhi.

Nirvitarka Samadhi: A Yogi enters into this Samadhi when he is able to concentrate on the **gross object without the awareness of its name, form and knowledge**.

It is the uncontrolled function of memory that leads to mental imagination. Memory links an object with time and other associations. By attaining success in Savitarka, a Yogi is able to enter into Meditation on an object without being distracted by its name and associated ideas. At this stage the mind, being pure like a crystal, so colored by the object, as if devoid of itself (*Y.S.*, I.43).

Savichara Samadhi: As *sattva* increases in the mind of a Yogi, he goes beyond the gross plane, and is able to meditate upon a **subtle element**. He reaches Savichara Samadhi by meditating upon a subtle element of space, air, fire, water or earth **with awareness of its name, form and knowledge.** Meditative reflection is its special feature (*Y.S.*, I.44).

Nirvichara Samadhi: A Yogi attains this Samadhi by meditating upon a **subtle element without awareness of its name, form and knowledge**. Then his mind identifies itself with the object without any distortion caused by imagination or memory (*Y.S.*, I.44).

Sananda Samadhi: When Nirvichara Samadhi intensifies, the Yogi ascends the ladder to subtler planes of existence. He reaches the level of **ego principle** or the plane of **bliss**. This Samadhi is called Sananda or blissful Samadhi, because the Yogi experiences blissful feeling. Meditation on bliss is free from deliberation and analysis. The body becomes calm when a feeling of bliss pervades the body.

Sasmita Samadhi: With the increasing purity of the mind, the Yogi goes beyond the ego center, and discovers the ego's causal plane. This is Sasmita Samadhi, i.e., Samadhi pertaining to the source of ego, the **Cosmic Mind** (*Mahat*). This is a state of **pure I-sense** (*Asmita*). In this state the mind becomes luminous, because the impurities of *rajas* and *tamas* have been removed.

Subtle object extends up to the unmanifested Prakriti. There is no other subtle object beyond Prakriti (*Y.S.*, I.45).

The **above six kinds of lower Samadhi** need a **support** or **basis** in the form of a gross, subtle, subtler or subtlest object on which the mind is concentrated. They are known as ***Sabija*** (or with seed) (*Y.S.*, I.46). In these six kinds of Samprajnata Samadhi, the mind is not entirely arrested. Therefore they require a basis or object of Concentration. Further the seeds of desire and attachment still remains in the mind. They can germinate at any time.

As the Yogi ascends the rungs of lower forms of Samadhi his intellect becomes exceedingly purified. He gets **intuitive vision,** which reveals that the Self is distinct from the mind (*Y.S.*, I.47). The intuitive vision reveals the Truth of the Self without any distortion. The light of intuition dispels all illusions. The **knowledge** that is acquired in this state is unique. It is *Ritambhara* (**filled with Truth**). It bears truth only and nothing else (*Y.S.*, I.48).

The **intuitive knowledge** (*Ritambhara Prajna*) surpasses all knowledge of the world. It is totally different from the knowledge gained by hearing or study of scriptures and inference, because it is a **specialized knowledge** (*Y.S.*, I.49). From scriptures and inference, only general conclusions can be arrived at. Specific aspects cannot be known. For example when we see smoke we know that there is fire, but the particulars regarding the nature and cause of fire cannot be understood from that. The scriptures teach us about the existence and glory of God. But we cannot claim to know God. God can be realized only when we transcend the mind, because the mind is matter, and matter cannot understand something subtler than itself. We can realize God only when we transcend the mind through non-reflective Samadhi. For that, the mind must be completely silent. That is why in Hindu mythology there is one form of God called Dakshnamoorti. Four learned persons had read all *Vedas* and *Upanishads* and heard all that was to be heard, but they still could not realize the truth. So they came to Dakshinamoorti and requested him to explain the highest Brahman. He just sat there in silence. After a while they got up, bowed down and said, "Swamiji, we have understood." And they went away, because only in silence it can be explained.

The intuitive knowledge is beyond the mind and it reveals the truth. What is the effect of intuitive knowledge?

Asamprajnata Samadhi: The **sublime impressions of the intuitive knowledge** are formed in the mind of the illumined Yogi. They **wipe out the latent impressions that have been caused by the worldly life** (*Y.S.,* I.50). They cause the destruction of the causes for afflictions, and so they make the mind to lose the capacity to produce its effects. The **activity of the mind ceases.**

The intuitive knowledge culminates in detachment. The Yogi's **discriminative knowledge** (*viveka khyati*) leads to **supreme dispassion** (*Para vairagya*). The Yogi, then, realizes that even the mind is no longer needed. He experiences controlled state of the mind. He begins to experience **Asamprajnata Samadhi** (Samadhi without seed), the **highest Samadhi** (Y.S., I.51).

This Samadhi puts an end to the seed of the worldly process. The impressions of the intuitive knowledge are **automatically effaced**, just as the automatic extinguishment of a fire after it has consumed its fuel. The mind now merges into its source, Prakriti. The Yogi attains **Liberation** from the fetters of karmas and from the cycle of births and deaths. He is **established in his own Self or Superconsciousness**. It may take one life or many lives for reaching this goal, depending upon the progress accomplished by the Yogi.

The Methods of reaching the highest Samadhi

***Videhas*:** Yogis who die after attaining the level of Sananda Samadhi are called *videhas* (those without body). They have destroyed their identification with the body, but have not yet attained the full Self-realization.. After death, they enjoy boundless bliss in the plane of pure mind, and are reborn in a family of Yogis, where they enter into Asamprajnata Samadhi with out much effort (*Y.S.*, I.19).

Prakritilayas: The same applies to *Prakritlayas* – Yogis who have attained Sasmita Samadhi, but did not go further. After death, they merge themselves into Prakrti and enjoy the bliss of liberated souls for certain period of time. But they also must be reborn in order to attain full Liberation (*Y.S.*, I.19).

Other Yogis: Others attain the highest Samadhi gradually by the **practice of Meditation with reverential faith, energy, memory, concentration and discriminative knowledge** (Y.S., I. 20). **Sustained interest and faith** gives yogi **energy** and discriminative knowledge. **Energy** brings him **memory**, which makes the mind undisturbed and collected and conducive to concentration. The **concentrated mind** attains **discriminative**

discernment. By the **discriminative enlightenment** or knowledge the Yogi **understands the real nature of things**. By retaining such knowledge and by cultivating **detachment** to the objects of the world he **attains Asamprajnata Samadhi** (*Y.S.*, I.20).

The above Yogis are of **nine kinds** according to the speed of practice – slow, moderate or fast. These speeds again are of three degrees each, viz., gentle intensity, moderate intensity or keen intensity. Among the followers of speedy methods, Yogis with intense intensity achieve Samadhi and the result thereof quickly (Y.S., I.21). When a Yogi, endowed with dispassion, keen enthusiasm and vigor, engages himself intensively in the practice of Yoga for attaining Liberation, he acquires momentum as he advances.

The time required for success again depends upon the degree of practice, viz., mild, medium or intense. On account of this difference, the speed of attainment of Samadhi and its result may be slower, faster or fastest (Y.S., I.22).

Is there **any other means** to attain the highest Samadhi? "**Yes**," says Patanjali, "There is another means, i.e., **by surrendering to God.**

Self-surrender to God

By intense devotion and surrendering oneself to God, one attains, by His Grace, Samadhi or absorption and attains Liberation. Self-surrender means dedication of everything to God and doing everything as an instrument of God, without any personal motive or interest. The devotee intensely feels in the heart of his hearts that it is God that is doing everything and that everything belongs to Him. There is no "I" in his case. This frame of mind banishes all egoistic feelings and only the thought of God ever remains in his mind (*Y.S.*, I.23). Lord Krishna says in the *Bhagavad Gita*,

> "To those who worship Me (God), renouncing all actions, in Me, regarding Me as the supreme goal, meditating on Me, with single-minded Yoga, and whose minds are fixed on Me, O Arjuna, verily I become ere long the savior out of the this mortal worldly life (12.6-7).

"Whatever you do, whatever you eat, whatever you offer in sacrifice, whatever you give , whatever you practice as austerity, O Arjuna, do it as an offering unto Me.
Thus shall you be freed from the bonds of actions yielding good and evil fruits; with the mind steadfast in the Yoga of renunciation, and liberated, you shall come unto Me" (9.27, 28).

The devotees who surrender themselves to God and fix their minds on God **attain Liberation** from the fetters of worldly life, and attain **immortality**.

Thiruvalluvar, a great saintly poet of Tamil Nadu, Southern India, in his sacred work *Thirukkural,* says,

Aravaazhi anthanan thaalsayrinthark kallaal
Piravaazhi neenthal arithu (Kural, 8)
Piravi perunkatal neenthuvar neenthaar
Iraivanati sayraa thaar (Kural, 10).

They alone cross the life's other oceans, who take refuge
at the Feet of One who is the ocean of Virtue (8).
Only those who reach the feet of God will cross the
Ocean of birth, others will not (kural 10).

In these two couplets Thiruvalluvar affirms that only those who surrender to God will cross the ocean of birth and death. (*See* also, last paragraph in 2.9 God (Isvara), above).

The States of mind and Samadhi

Patanjali describes the manner in which the mind undergoes modifications during the attainment of Samadhi as follows:

Arrested state: The ordinary state of mind is restlessness. In this state the mind is dominated by *rajas* and *tamas*. It is attracted by sense-objects. It jumps from one thing to another like a restless monkey. By the practice of *Yamas, Niyamas* and other stages of Yoga, the mind is purified and reaches an arrested state when Samadhi is attained. As the mind is made up of three *gunas*, which are always changing, changes takes place even in the arrested state. But there is no manifestation of that change, as it cannot be perceived by the mind. What is the nature of that change?

In the arrested state, the mind is blank. During the practice of Asamprajnata or Nirodha Samadhi, the impressions of outgoing modifications (*Vyuthama Samskaras*) gradually decrease and the impressions of control (*Nirodha Samskaras*) increase. This is unnoticed and not perceived like the knowledge of an object. Although not perceptible, it is in effect a change. The mind is then in a moment of **blankness**. This is the change of increasing the state of stoppage of all mental states. The characteristic of the arrested state of mind is that in every moment there is destruction of latent impressions of modifications and development of latent impressions of control (*Y.S.,* III.9).

Tranquil state: When the impressions of control overcome the impressions of outgoing modifications, proficiency is attained in the art of keeping the mind in an arrested state. Then the mind attains a continuous undisturbed or concentrated state. This is known as **tranquil state** (*Prashanta vahita*) (*Y.S.*, III.10).

Samadhi state: Mind has two characteristics. The first is all-sided distraction, taking up various objects or indulging in all kinds of desires and thinking of the past and the future. The second is one-pointed or focused state, fixing the attention on one object. When all-sided distraction disappears and the **one-pointedness emerges**, then the mind gets **Samadhi state** (*Samadhi Parinama*) (*Y.S.*, III.11).

One-pointed or Stabilized state (*Ekagrata Parinama*): Then, as the rising and subsiding of the thought-waves or the past and the present modifications become similar, the mind stabilizes in one-pointedness. It refers to appearance and disappearance of the same idea. The mind is then **habitually one-pointed**. The idea of time will vanish. In everyday life we see that when we are absorbed in reading a book, we do not note the time at all (*Y.S.*, III.12).

The difference between the above three states of Concentration mentioned in *Sutras* III.9, 10, and 12 is as follows:

In the **first**, the disturbed impressions of the out-going modifications are **merely held back**, but not altogether erased by the impressions of control, which have just come in; in the **second**, the disturbed impressions are **completely suppressed** by the impressions of control, which become dominant; in the **third**, which is the highest, there is no question of suppression but only similar impressions succeed each other in a stream like the oil poured from one vessel to another, i.e., the **concentration becomes habitual**.

Changes in the Mind, Elements and Senses

The **three changes** or transformations mentioned in the above *Sutras* III.9, 11 and 12 refer to the **essential attribute** (*Dharma*), temporal character or **periods of time** (*Lakshana*) and **condition** (new or old) (*Avastha*). Just as these are changes in the conditions of the mind, so there are **similar** three kinds of **changes in the elements and senses**. This is how we speak of the differences between objects.

Essential attribute: This refers to the essential nature or **form**. When one characteristic disappears and another rises, that is called a change of attribute. Suppose there is a lump of earth. When it is transformed into a pot, it gives up the form of lump and takes that of pot. Similarly, mind changes into thought waves or modifications. This is a transformation of form.

Temporal change: This refers to change in relation to the three periods of **time** – past, present and future. For example, before the pot was made out of a lump of clay, the pot is in the future; when it has been made, it is in the present; when it has been destroyed, it is in the past. Similarly when the mind passes through past, present and future moments of time, it is transformation as to time.

Condition: This refers to change in the **present condition**. For example a pot is first called new. After some time it is called old. When the disturbing impressions of the mind are strong, and the controlling impressions are weak, and vice versa, it is transformation as to condition (*Y.S.*, 13).

All the entities except the Purusha or Self are subject to the threefold transformation mentioned above.

A yogi has voluntary control over the transformations of the mind through the practice of Concentration, Meditation and Samadhi.

Potentiality for Changes

Every object is capable of undergoing a series of changes with various forms or characteristics. Behind the changes there is a substratum or basic stuff, which is constant. Its various forms or characteristics were in the state of indescribability before their appearance. Then the characteristics or forms were hidden in the basic stuff and will arise in proper conditions. When a particular characteristic arises, it is present. When it dies out, it is known as past.

For example, water is a basic stuff out of which different forms such as ice, iceberg, snow and vapor may arise, and the possibility of assuming one of these forms is hidden in it. When the water is converted into ice, the basic water remains in its own nature. When the ice melts back into water, it still remains the same. Similarly there are constant changes and variations in the external objects of the world (*Y.S.*, III.14).

Succession of Changes

The succession or sequence of changes is the cause of manifold evolution or modification. For example, dust, clod, pot, bits are sequences of earth. The form, which follows another is its sequence. Clod disappears and pot appears. This is sequence of change of form. The appearance of a pot from its potential state represents temporal transition from the unmanifested to its present state, while the disappearance of the clod of earth represents a temporal transition from the present state to the past. Sequence of change of condition is also similar. A new pot becomes old in course of time. The oldness is only the result of the sequence of change taking place every moment.

A mind has two characteristics, viz., patent and latent. The patent characteristics are perceivable ones, e.g., cognition or feeling. The latent ones merely exist as subconscious, e.g., arrested state, latent impressions, etc (*Y.S.*, III.15).

Samyama

The **combined practice** of Concentration, Meditation and Samadhi on the same object is called Samyama (*Y.S.*, III.4). Samyama is the Yogic technical name for this practice. It is a means of acquiring knowledge in respect of, and control over, the object contemplated upon. The mind is directed to a particular object and fixed on it. Concentration takes place on all aspects of the object. In one Samyama there can be several chains of Concentration, Meditation and Samadhi. Sustained knowledge of the object is acquired by repeatedly practicing all the three.

By **mastering Samyama**, a Yogi attains the **light of knowledge** or insight (*prajna*) (*Y.S.*, III.5). As a result of constant practice of Samyama, his intellect becomes free from the influence of *rajas* and *tamas*. His reasoning faculty becomes luminous and bright with the light of intuitive knowledge. This **light** can be used to unfold various **psychic powers**.

However, the Yogi should not **be led astray** by the love of powers, but he must endeavor to go to the very **source of powers, the Self** within for attaining Liberation or perfect freedom. That is the true purpose of Yoga.

Samyama is to be applied to the different stages of practice: first to the lower stages, after mastering lower stages, next to intermediate stages, and finally to higher stages. One who has attained a higher stage by God's grace (earned by intense devotion) need not practice Samyama in respect of the lower stages (*Y.S.*, III.6).

The **stages** or planes to which Samyama to be applied are:

1. **Gross stage**: In this stage the Yogi experiences Savitarka Samadhi and Nivitarka Samadhi.
2. **Subtle stage**: In this he experiences Savichara Samadhi and Nirvichara Samadhi.
3. **Subtler stage**: At this stage the Yogi experiences Sananda Samadhi.
4. **Subtlest stage**: Here Sasmita Samadhi is experienced, and finally the Yogi transcends all these four planes of Prakriti through the practice of Asamprajnata Samadhi.

With reference to the lower Samadhi—Samprajnata—the first five steps of Raja Yoga are the external means, while Samyama (the last three steps) is internal means (*Y.S.*, III.7). But with reference to the highest Asamprajnata (Seedless) Samadhi, Samyama is the external means, while supreme dispassion is the internal means, because seedlessness is attained when Concentration, Meditation and Samadhi are also absent (*Y.S.*, III.8).

Patanjali describes the various powers that are attained by practicing Samyama on the three transformations or changes of primary characteristics, temporal character and condition (described in *Sutra* III. 13, *see* paragraph **"Changes in the mind, elements and senses,"** above) with reference to any object (*Y.S.*, III.16-55).

The **occult powers** are **not relevant** to the **true purpose of Yoga**. They invariably intensify worldliness and, in the end intensify suffering. Though Patanjali as a scientist points out the possibilities of his science, he does not miss an opportunity to **warn** the Yogis **against these powers**.

If a Yogi is overtaken by the desire to acquire through Samyama various powers and their enjoyment, he is drifted away from the goal of Yoga. After all, the powers are mere manifestations. As Swami Vivekananda says, "They are no better than dreams. Even omnipotence is a dream. It depends on the mind. So long as there is mind, it can be omnipotent; but the goal is beyond even the mind."

Even the mastery over Nature and omniscience are the results of the lower Samadhi. But as the Yogi advances beyond the lower Samadhi into Asamprajnata Samadhi, the highest Samadhi, he develops **dispassion** even **toward these powers** (*siddhis*). He destroys the very root of evil in the form of ignorance. The mind ceases to be operative and he goes beyond supernormal experiences such as omniscience, etc. The Yogi realizes the intrinsic freedom of the Self, and is **established in Absolute Existence**, peace and bliss (*Y.S.*, III.50).

Questions

1. What is Samadhi? How is it attained?

2. How is Samadhi classified?
3. What is Samprajnata Samadhi?
4. Describe the nature of lower forms of Samadhi with seed.
5. Explain the nature of intuitive knowledge.
6. What is Asamprajnata Samadhi? What is its nature?
7. Differentiate Asamprajnata Samadhi from Samprajnata Samadhi.
8. Explain self-surrender to God as an alternate means to attaining the highest Samadhi.
9. Describe the different modifications undergone by the mind during Samadhi.
10. Explain potentiality for changes and succession of changes of objects.
11. What is Samyama? Can it be applied straight away at the higher stage?
12. Are the occult powers relevant to the true purpose of Yoga? Give reasons for your answer.

3.11: Liberation (Kaivalya)

Introduction

Liberation or Self-realization means freedom from karmic bondage and suffering. This is the **supreme goal of life**. This is the **final stage of Yoga**. In this state a Yogi realizes his true nature, realizes that he is the eternal **Self** or Atman. He abides in it. How does this supreme state of Superconsciousness or a state of blissful peace reached?

Discriminative Knowledge (*Viveka Khyati*)

When a Yogi advances in the stage of Sasmita Samadhi, the highest limit of lower Samadhi, he experiences discriminative knowledge or discernment. He becomes disinterested even in omniscience (*Prasamkhyana*), the highest state of elevation, resulting from discriminative enlightenment. When this is maintained ceaselessly, the Yogi develops **supreme dispassion** (*Para vairagya*). Then he begins to see his Self as distinct from his mind, and therefore, he develops a mystic inclination to turn away from the mind itself.

This state has **two aspects**. **First**, the flow of discriminative knowledge becomes firm. **Second,** Asamprajnata (or Nirodha) Samadhi is developed. The seeds of all latent impressions are destroyed. This state is figuratively described as ***Dharma Megha*** (Cloud of virtue), which rains down the bliss of Liberation. The Yogi, who is dispassionate even towards discriminative enlightenment, attains *Dharma Megha* Samadhi (*Y.S.*, IV.29)

Jivanmukta

As a result of *Dharma Megha* Samadhi, the five causes of afflictions (*klesas*) (ignorance, egoism, etc.) and their effects and the three types of karmas (good, bad and mixed) are destroyed. The Yogi becomes established in the Self, i.e., he is liberated while alive (*Jivanmukta*) and becomes free from sorrow and the cycle of birth and death. His Consciousness merges in the Light of Reality (*Y.S.*, IV.30).

A *Jivanmukta* is free from egoism, beyond the three *gunas* of nature, and free from desires and selfishness. He is established in the Self. He sees his Self in all beings and all beings in his Self. He is not bound by actions, for he has no sense of doership and any personal motive. He just performs actions as an instrument of God. He is beyond pleasure and pain and other dualities. He renders service for the welfare and spiritual development of people, and his service is God's service. This is purely a selfless service. Sri Ramakrishna gives a beautiful parable about this.

Once, a few people went to a garden, having been told that there were beautiful big trees there. But the garden was completely surrounded by high walls, and they couldn't even see what was inside. With great effort one person managed to climb the wall and see inside. He saw such luscious fruit that immediately he jumped in. Another person climbed up and immediately jumped in the same way. Finally, a third person climbed up, but when he saw it, he said, "My God, how can I jump in now? There are so many hungry people below who don't know what is here or how to climb up." So, he sat on the wall and said, "Hey, there are a lot of fruits, come on. If you try hard you can come up like me." He lent a hand, pulling people in. Such persons are teachers. They guide others to attain Liberation. Their service is selfless and it is a service to God.

Limitless Knowledge

Rajas and *Tamas* are coverings over knowledge. As a result of *Dharma Megha* Samadhi the veils of illusion are lifted and the **knowledge** of the Yogi becomes **limitless** (*Y.S.*, IV.31). Knowledge becomes limitless when perfect calmness is achieved. Before he reached this state of illumination, his knowledge was conditioned by ignorance. During the state of ignorance, things to be known or the objects of the world were limitless before his vision. Now the Yogi has ascended the height of wisdom, the world of things to be known becomes insignificant like fireflies in the limitless space. In other words the relative world process is transcended, and the Yogi has nothing more to know. He becomes knowledge itself.

What happens to *Gunas*?

The three *gunas* of Prakriti operate for the sake of souls (Purushas). Through their modifications the *gunas* provide experience of pleasure and pain in order to lead the Purushas to the state of Liberation. On the attainment of *Dharma Megha* Samadhi and Liberation by a Yogi the three *gunas*, having fulfilled their purpose for him, cease to have any further series of modifications. With the fulfillment of their twofold purpose, viz., giving experience and liberation to the Yogi, the contact of the Prakriti ends, and the sequence of modifications of the *gunas* ceases to manifest in him. That is, although these *gunas* continue to operate for the unenlightened souls, the liberated Yogi becomes free from them (*Y.S.*, IV.32). Then what does happen to Prakriti? Swami Satchidananda, in his commentary, gives an analogy to explain it. Imagine a mother with a number of children. They all go out to play and get dirty. When they return, totally covered with dirt, she puts them into the bathtub and turns on the shower. Of course, she can't clean everyone at the same time, so she washes one by one. Once a baby is clean, she takes it out of the tub and say, "Go, get dry and hop into bed." Will she stop working? No. There are still more dirty children in the tub. Mother Prakriti is just like that. She stops functioning with the liberated soul. But she has a lot of work to do with others.

Sequence is of the nature of incessant flow of moments, and it is conceived only when a change becomes noticeable. For example a new piece of cloth does not become old unless it has passed through the sequence of moments, which has caused its change, though not noticeable at that time. The essence of Purusha and the *gunas* is their eternality. Though the *gunas* change or mutate, their essential nature never changes or destroyed. They are changeably eternal. Therefore, their changeability never comes to an end; but in the various evolutes like intellect through which the *gunas* manifest themselves, the sequence of moments comes to an end. But in the *gunas*, which are eternal, the sequence never ends, and is experienced by other persons still in bondage (*Y.S.*, IV.33).

As the *gunas* have no further service to render to the liberated Yogi, they return to their source, Prakriti. Then the Yogi abides in his Self. In other words, the Supreme Consciousness or Self just exists. It remains alone all time (*Y.S.,* IV.34). The liberated Yogi becomes free from suffering, free from bondage of karma, free from the cycle of birth and a death. After he leaves his body he is not born again. He attains immortality. His soul merges into the Cosmic Self or Brahman.

The nature of the state of Liberation is beyond description and our comprehension. Patanjali, knowing the futility of such a task, has not even attempted it.

The liberated one realizes the Light of lights. The *Mundaka Upanishad* describes this state as:

> "The sun shines not there, nor the moon and stars; there lightning shines not, where then could the fire be? Everything shines only after that shining Light. This Light illumines all the whole world." (2.2.11)
>
> "...One's deeds and the Self, consisting of intellect, all become one in the Supreme immutable Being" (3.2.7).
>
> "Just as the flowing rivers disappear in the ocean, casting off name and form, so, the knower, freed from name and form, attains to the Divine person, higher than the high" (3.2.8).

Questions

1. What is *Dharma Megha* Samadhi? How is it attained? What are its results?
2. What is Liberation? What are its effect?
3. What does happen to the three *gunas* when a Yogi attains Liberation?

Bibliography

Rukmani, T. S., *Yoga Sutras of Patanjali*, Montreal: Chair in Hindu Studies, Concordia University, 2001.

Swami Hariharananda Aranya, *Yoga Philosophy of Patanjali*, Albany: State University of New York Press, 1985.

Swami Jyotirmayananda, *Raja Yoga Sutras*, Miami: Yoga Research Foundation, 1978.

Swami Prabhavananda and Isherwood, Christopher, *How to Know God – the Yoga Aphorisms of Patanjali*, Hollywood: Vedanta Press, 1981.

Swami Satchidananda, *The Yoga Sutras of Patanjali*, Yogaville, VA: Integral Yoga Publications, 1990.

Swami Vivekananda, *Raja Yoga*, New York: Ramakrishna-Vivekananda Center, 1982, pp. 94-221.

Taimni, I. K., *The Science of Yoga*, Wheaton, Illinois: The Theosophical Publishing House, 1967.

References

1. The *Bhagavad Gita* is one of the ancient scriptures of India; it primarily deals with Yoga. For further details, *see* **Yoga in Epics** in 1. 2 Yoga Tradition in India, below.
2. Sri Ramakrishna (1836-1886) was the God-incarnate of India. In his life he has demonstrated the unity of all Faiths. The Ramakrishna Mission established by his great disciple Swami Vivekananda propagates his teachings.
3. *Isa Upanishad* is one of the earliest *Upanishads*, the ancient scriptures, composed before five thousand years in India. The *Upanishads* contain eternal truths relevant to all ages and all people. For details, *see* Yoga in Upanishads in 1.2 Yoga Tradition in India, below.
4. The *Bible* consists of Old Testament and New Testament. The Old Testament is the holy scripture of Jewish people. The New Testament is the scripture of Christianity.
5. *Siva* is one of the Hindu names of God.
6. Shankaracharya (eight century AD) was one of the greatest sages of India. He expounded the non-dualistic Vedanta. He wrote commen-

taries on eleven *Upanishads* and other ancient Scriptures.

7. Classical Yoga is Yoga as codified by the sage Patanjali. Yoga of the period prior to this codification is called Pre-classical Yoga.
8. A great Sage lived in Tiruvannamalai in Southern India during twentieth century. He attained Super- consciousness through Jnana Yoga.
9. A Sage (twentieth century) of South India, who attained God-realization by constantly doing *Japa* on Lord Rama.
10. Swami Ramdas, *The Divine Life,* Anandashram: Anandashram, 1991, p. 250.